COLLEGE READINESS MATH REVIEW
FOR TSI, SAT AND ACT

BRENDA VOYLES

authorHOUSE

AuthorHouse™
1663 Liberty Drive
Bloomington, IN 47403
www.authorhouse.com
Phone: 833-262-8899

Published by AuthorHouse 08/06/2020

ISBN: 978-1-7283-6879-5 (sc)
ISBN: 978-1-7283-7020-0 (e)

Library of Congress Control Number: 2020915103

Table of Contents

PLACE VALUE

*A decimal seperates whole numbers from pieces of a whole. Every place value to the left of the ones place value is 10 times larger and every place value to the right of the ones place value is 10 times smaller.

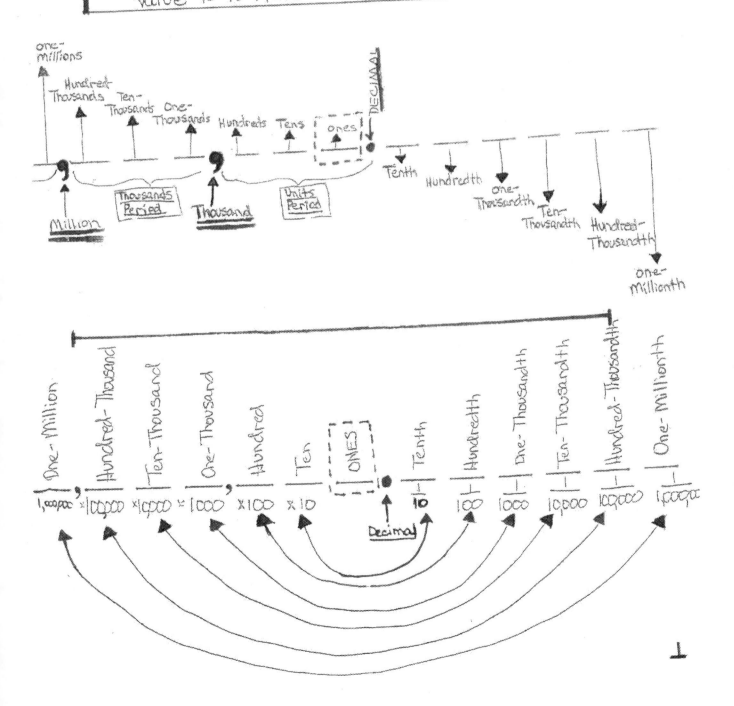

COMPARING DECIMALS

* Decimals can only be compared by lining up place values and compare each place value from left to right — line up decimals.

Step 1. Line up decimals and each place value to left and right.

2. Start with place value fartherest to left and move to right if needed

Example 1 — List these numbers from least to greatest

1.03, 0.165, 0.23, 0.3

↓

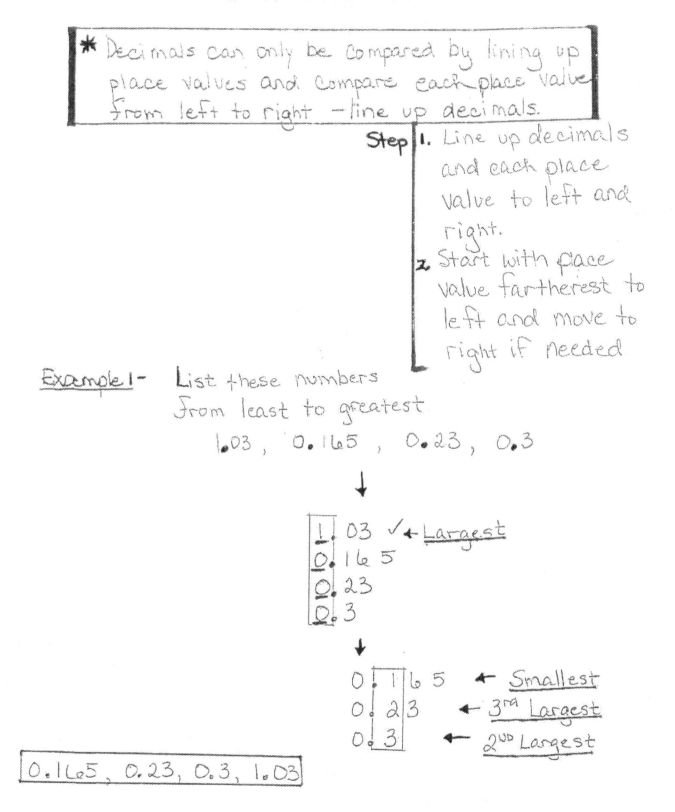

1.	03	✓ ← Largest
0.	165	
0.	23	
0.	3	

↓

0.	1	65	← Smallest
0.	2	3	← 3ʳᵈ Largest
0.	3		← 2ⁿᵈ Largest

0.165, 0.23, 0.3, 1.03

ADDING AND SUBTRACTING DECIMALS

* <u>Add and Subtract Decimals</u> by lining up decimals to line up like place values to add or subtract and drop decimal down into answer

Step 1. Line up decimals
2. Add or subtract
3. Drop decimal down into answer

Example 1 - <u>Add:</u> 41.06 + 8.2

$$
\begin{array}{r}
41.06 \\
8.2 \\
\hline
49.26
\end{array}
$$

49.26

Example 2 - Subtract: 26.5 - 14.63

Step 1. Line up decimals
2. When Subtracting, make sure all place values are full - add a <u>0</u> to any empty place values

$$
\begin{array}{r}
26.5 \\
-14.63 \\
\end{array}
$$

$$
\begin{array}{r}
26.50 \\
-14.63 \\
\end{array}
$$

$$
\begin{array}{r}
2\overset{5}{6}.\overset{14}{5}0 \\
-14.63 \\
\hline
11.87
\end{array}
$$

11.87

3

MULTIPLYING DECIMALS

---

* Decimals <u>do</u> <u>not</u> have to be lined up to multiply. Multiply numbers **ignoring** decimals. Once multiplied, **add all** place values to <u>right</u> of decimals in the original problem. Then decimal should be moved that number of place values to the <u>left</u>, starting from the right **end of the answer.**

Example 1:   **Multiply:**

$$.16 \times 34.6$$

↓

$$\overset{2}{3}\overset{3}{4}.6$$
$$\underline{\quad .1\,\cancel{6}}$$

$$2\ 0\ 7\ 6$$
$$3\ 4\ 6\ \cancel{8}$$
$$\overline{5\ 5\ 3\ 6}$$

↓

Step 1. Easiest to put number with most digits on top to multiply - <u>Do</u> <u>not</u> line up decimals

2. Multiply as usual as if there were no decimals

3. Count the number of place values to <u>right</u> of decimals in problem

4. In answer, start at <u>right end</u> and move decimal to <u>left</u> that number of place values

$$\left.\begin{array}{l} 3\,4\,.\,\underline{6} \\ \ .1\ \underline{6} \end{array}\right\}$$ <u>3 place values to right of decimals</u>

$$5\underset{\curvearrowleft}{.}5\underset{\curvearrowleft}{3}\underset{\curvearrowleft}{6}\,\times$$ ← <u>Start here and move 3 place values to left</u>

↓

$$\boxed{5.536}$$

4

DIVIDING DECIMALS

* Decimals may be inside, but are not allowed outside of division sign to divide. Any decimal outside the division sign (divisor) must be moved to the right (end of the number) next to the division sign. Repeat same number of moves to the right of the number inside division sign - add 0 if any missing place values. IF no decimal outside division sign, pull decimal up to the top of division sign.

Step 1. IF no decimal outside division sign, pull decimal inside division sign up into answer
2. Divide as usual

Example 1 - Divide: $50.4 \div 42$

↓

$$42\overline{\smash{)}50\overset{\bullet}{.}4}$$

↓

$$\begin{array}{r} 1.2 \\ 42\overline{\smash{)}50.4} \\ \underline{42}\downarrow \\ 84 \\ 84 \end{array}$$

$\boxed{1.2}$

Example 2- Divide: 58.9 by .19

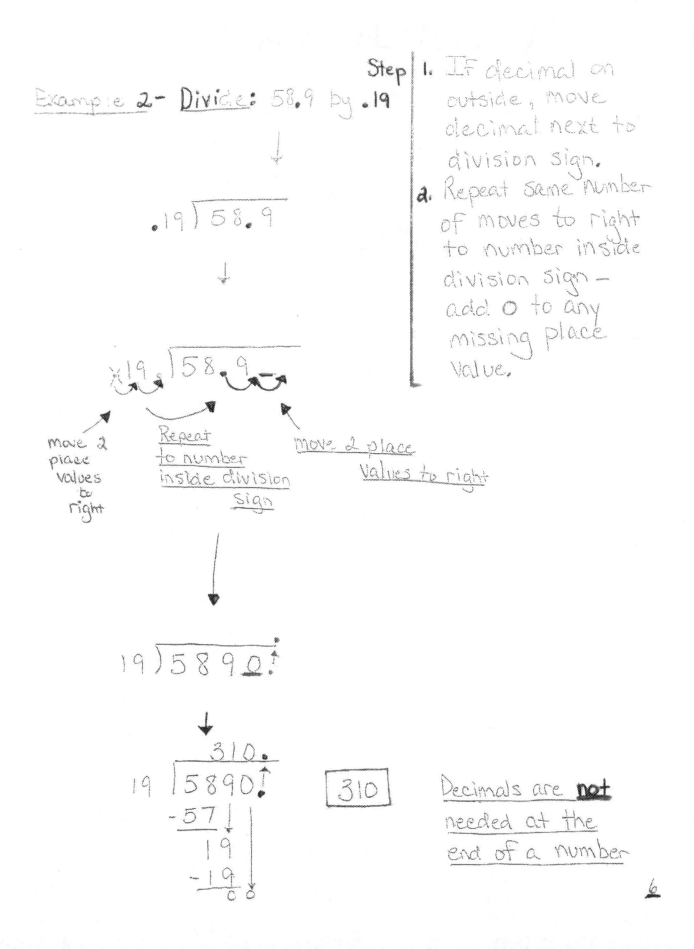

$$.19 \overline{)58.9}$$

1. IF decimal on outside, move decimal next to division sign.

2. Repeat same number of moves to right to number inside division sign — add 0 to any missing place value.

move 2 place values to right

Repeat to number inside division sign

move 2 place values to right

$$19 \overline{)5890.}$$

$$\begin{array}{r} 310. \\ 19 \overline{)5890.} \\ -57 \\ 19 \\ -19 \\ 00 \end{array}$$

310

Decimals are **not** needed at the end of a number

6

SCIENTIFIC NOTATION

* <u>Scientific Notation</u> is a mathematical expression used to represent a decimal number between 1-10 multiplied by a base of 10, so you can write a large or small number using less digits.

Step 1. Find and move decimal (if no decimal, it is at the back end of the number) so that the number is expressed as a number between 1-10

2. Count number of place values the decimal moved. This number is the exponent with a base of 10

Example1- Write in scientific notation

76,201

↓

76,201. ← Identify decimal

↓

7.6 2 0 1 ✗ ← Decimal moved 4 place values to left

↓

7.6 201

↓

7.6201 × 10⁴ ← 4 because decimal moved 4 place values to left

$$7.6201 \times 10^4$$

Positive exponent because original number was bigger than 10

7

Example 2 - Write in scientific notation

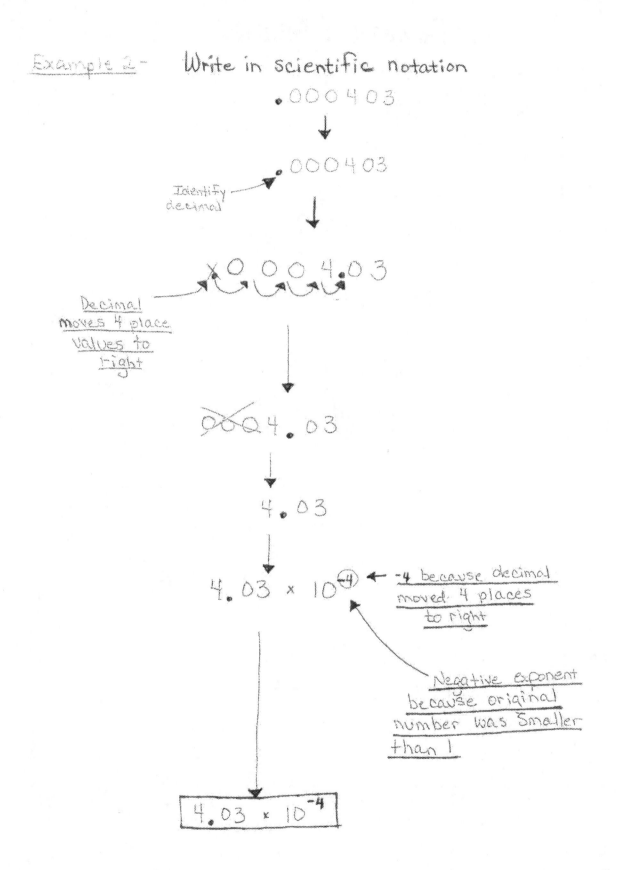

.000403

↓

.000403

Identify decimal

↓

x0 0 0 4.03

Decimal moves 4 place values to right

↓

000 4.03

↓

4.03

↓

4.03 × 10⁻⁴   ← -4 because decimal moved 4 places to right

Negative exponent because original number was smaller than 1

4.03 × 10⁻⁴

8

GREATEST COMMON FACTOR (GCF)

* <u>Factors</u> **are** numbers that can be evenly divided into a number without remainder. To find factors, always start with 1 and try dividing by every number until the next or same number is in the second column

Step | 1. Start with 1 as first factor

2. Try every number until the next or same number is in the second column

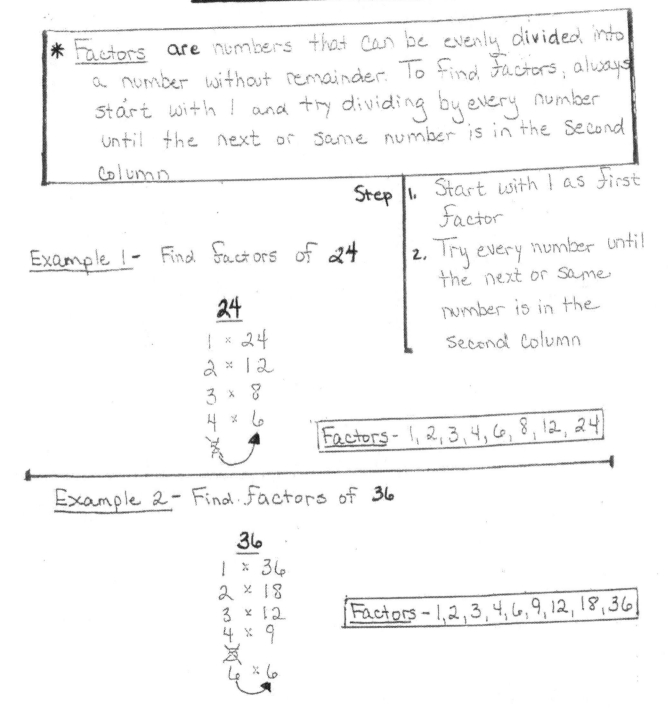

Example 1- Find factors of **24**

<u>24</u>

1 × 24
2 × 12
3 × 8
4 × 6

Factors- 1, 2, 3, 4, 6, 8, 12, 24

Example 2- Find factors of **36**

<u>36</u>

1 × 36
2 × 18
3 × 12
4 × 9
6 × 6

Factors- 1, 2, 3, 4, 6, 9, 12, 18, 36

* <u>Greatest Common factor</u> (GCF) - the biggest number that all numbers have in common (used to simplify fractions and simplify polynomials)

<u>Example 3</u> - Find <u>greatest common factor</u> of 24 and 36

24 - 1, 2, 3, 4, 6, 8, 12, 24

36 - 1, 2, 3, 4, 6, 9, 12, 18, 36

GCF

GCF = 12   <u>Common factors</u> - 1, 2, 3, 4, 6

LEAST COMMON MULTIPLE (LCM)

* __Multiples__ are products found by multiplying a number by 1, 2, 3, ...

Step 1. Multiply a number by 1 (any number is a multiple of itself

2. Continue number by 2, 3, 4 ... (each product is a multiple)

Example 1 - Find first 5 multiples of 6

$6 \times 1 = 6$
$6 \times 2 = 12$
$6 \times 3 = 18$
$6 \times 4 = 24$
$6 \times 5 = 30$

Multiples - 6, 12, 18, 24, 30

* __Least Common Multiple__ (LCM) - The __Smallest__ multiple that 2 or more numbers have in Common (used to add and subtract fractions)

Example 2 - Find least common multiple of 6 and 8

$6 \times 1 = 6$ $8 \times 1 = 8$
$6 \times 2 = 12$ $8 \times 2 = 16$
$6 \times 3 = 18$ $8 \times 3 = \boxed{24}$
$6 \times 4 = \boxed{24}$

LCM = 24

11

Example 3 Find least common multiple (LCM) of 4, 8 and 10

$4 \times 1 = 4$ $8 \times 1 = 8$ $10 \times 1 = 10$

$4 \times 2 = 8$ $8 \times 2 = 16$ $10 \times 2 = 20$

$4 \times 3 = 12$ $8 \times 3 = 24$ $10 \times 3 = 30$

$4 \times 4 = 16$ $8 \times 4 = 32$ $10 \times 4 = \boxed{40}$

$4 \times 5 = 20$ $8 \times 5 = \boxed{40}$

$4 \times 6 = 24$

$4 \times 7 = 28$

$4 \times 8 = 32$

$4 \times 9 = 36$ $\boxed{LCM = 40}$

$4 \times 10 = \boxed{40}$

12

PROPER VS IMPROPER FRACTIONS

---

* <u>Proper Fraction</u> - Number greater than 0 but less than 1 - Numerator is smaller than Denominator

---

Example - $\div \dfrac{Numerator}{Denominator}$ $\dfrac{2}{3}$, $\dfrac{4}{5}$, $\dfrac{1}{2}$

---

* <u>Improper Fraction</u> - Number greater than 1 - Numerator is larger than Denominator (Top Heavy!)

Example - $\div \dfrac{Numerator}{Denominator}$ $\dfrac{3}{2}$, $\dfrac{5}{4}$, $\dfrac{2}{1}$

13

Converting Improper Fractions To Mix Numbers

---

*A mixed number is a whole number followed by a fraction

Step | 1. Divide numerator by Denominator
2. If there is any remainder, write as a ratio of remainder ÷ by divisor

Example 1 - Write $\frac{7}{2}$ as a mixed number -

$\underline{7}$ numerator
2 denominator

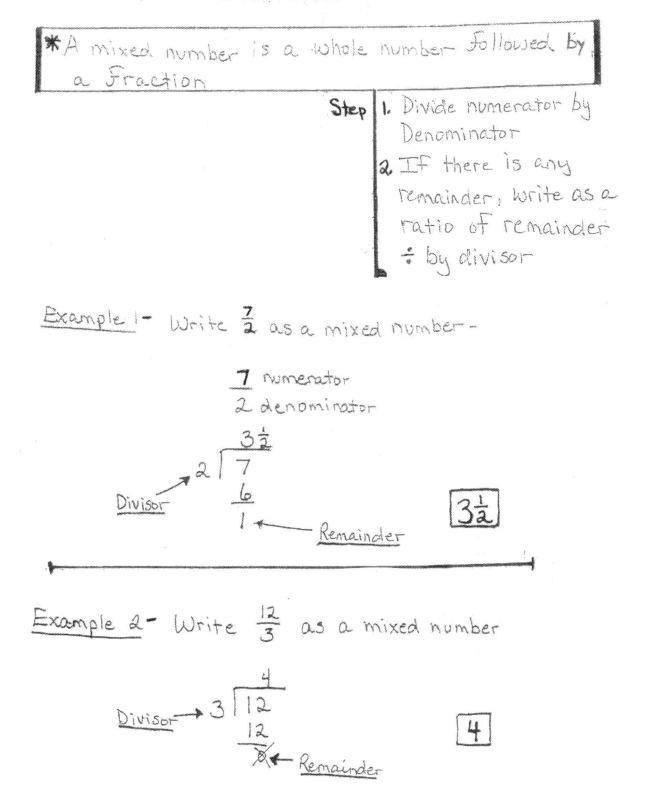

$$2\overline{)7}^{\,3\frac{1}{2}}$$

Divisor

7
6
1 ← Remainder

$\boxed{3\frac{1}{2}}$

Example 2 - Write $\frac{12}{3}$ as a mixed number

Divisor → $3\overline{)12}^{\,4}$
12
0 ← Remainder

$\boxed{4}$

14

CONVERTING MIXED NUMBERS TO IMPROPER FRACTIONS

* __Mixed Numbers__ must be written as improper fractions to multiply or divide fractions

Step
1. Multiply denominator by the whole number
2. Add this product to the numerator
3. Place this sum on the top of the fraction and keep the original denominator

Example 1- $4\frac{2}{3}$

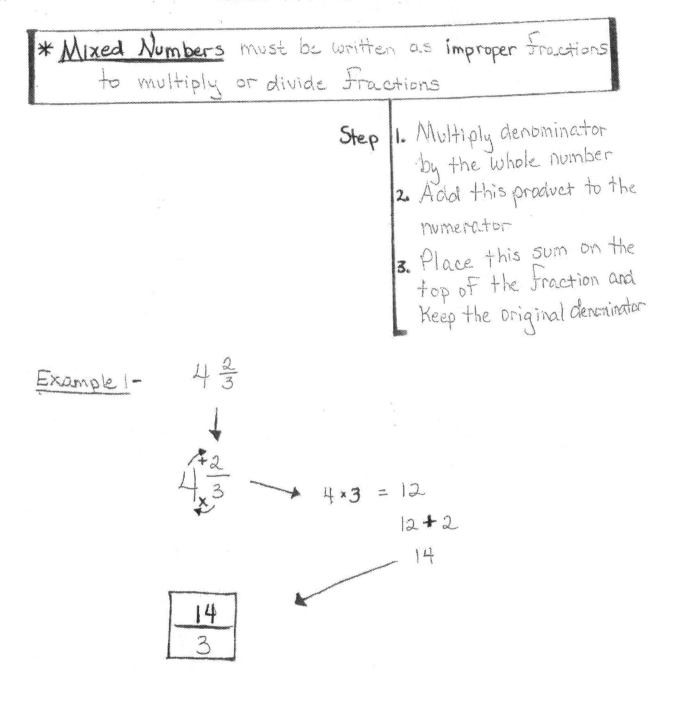

$4\times 3 = 12$

$12 + 2$

14

$$\frac{14}{3}$$

ADDING AND SUBTRACTING FRACTIONS

* To add or subtract fractions, <u>denominators</u> <u>must</u> be the same. If they are the same, keep denominator and add or subtract numerators. Reduce if possible.

Example 1 - $\frac{1}{6}$ + $\frac{2}{6}$ ← Same denominators

$\frac{3}{6}$ ← Keep denominator,
              add numerators

$\boxed{\frac{1}{2}}$ ← Reduce

Example 2 - $\frac{3}{4}$ - $\frac{1}{2}$ ← Find Least Common
                                          multiple for
  ↓   denominators
LCM of 4 and 2
     is $\underline{4}$

  ↓

$\frac{3}{4}$ - $\frac{2}{4}$ ← Rewrite with
                                  denominators
  ↓   of 4 (multiply
$\boxed{\frac{1}{4}}$ $\frac{1}{2} \cdot \frac{2}{2}$)

Example 3 - $\dfrac{3}{5} + \dfrac{2}{3}$.   ← Find least common multiple for denominators

↓

LCM of 5 and 3 is $\underline{15}$

↓

$\dfrac{9}{15} + \dfrac{10}{15}$ ← multiply $\dfrac{3}{5} \cdot \dfrac{3}{3}$ and

$\dfrac{2}{3} \cdot \dfrac{5}{5}$

↓

$\boxed{\dfrac{19}{15}}$ to get same denominators

⟵————————————————————⟶

* To add or subtract mixed number, rewrite as improper fractions. Then use adding and subtracting rules.

Example 4 - $2\dfrac{1}{3} + 2\dfrac{1}{4}$

↓

$\dfrac{7}{3} + \dfrac{9}{4}$ ← Write as improper fractions

↓

$\dfrac{28}{12} + \dfrac{27}{12}$ ← Rewrite with common denominators $\left(\dfrac{7}{3} \cdot \dfrac{4}{4} \qquad \dfrac{9}{4} \cdot \dfrac{3}{3} \right)$

↓

$\dfrac{55}{12}$

↓

$\boxed{4\dfrac{7}{12}}$ ← Write as mixed number

MULTIPLYING AND DIVIDING FRACTIONS

* To multiply fractions, cancel out any common factor in numerator and denominator __First__. Then, multiply numerators and denominators straight across.

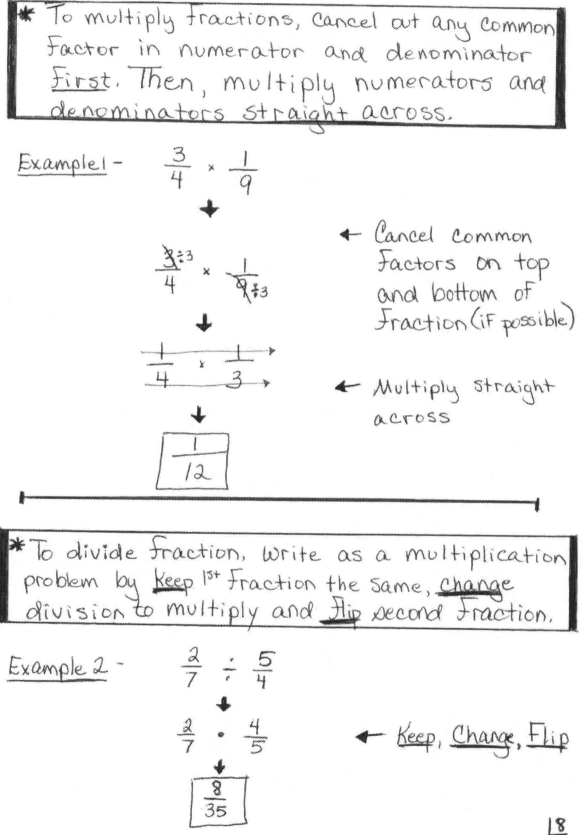

Example 1 - $\dfrac{3}{4} \times \dfrac{1}{9}$

$\dfrac{3 \div 3}{4} \times \dfrac{1}{9 \div 3}$

← Cancel common factors on top and bottom of fraction (if possible)

$\dfrac{1}{4} \times \dfrac{1}{3}$

← Multiply straight across

$\boxed{\dfrac{1}{12}}$

* To divide fraction, write as a multiplication problem by __Keep__ 1st fraction the same, __change__ division to multiply and __Flip__ second fraction.

Example 2 - $\dfrac{2}{7} \div \dfrac{5}{4}$

$\dfrac{2}{7} \cdot \dfrac{4}{5}$

← __Keep__, __Change__, __Flip__

$\boxed{\dfrac{8}{35}}$

18

* To multiply or divide mixed numbers, you __must__ rewrite all mixed number as improper fractions. Then follow rules for multiplying and dividing fractions

Example 3 - $2\frac{1}{3} \cdot 4$

⬇

$\frac{7}{3} \cdot \frac{4}{1}$ ← Rewrite as improper fractions

⬇

$\frac{28}{3}$ ← Multiply straight across (reduce if possible)

⬇

$\boxed{9\frac{1}{3}}$ ← Write as a mixed number

Example 4 - $2\frac{1}{3} \div 1\frac{1}{2}$

⬇

$\frac{7}{3} \div \frac{3}{2}$ ← Write as improper fractions

⬇

$\frac{7}{3} \cdot \frac{2}{3}$ ← __Keep__, __Change__, __Flip__

⬇

$\frac{14}{9}$ ← multiply across

⬇

$\boxed{1\frac{5}{9}}$ ← Write as mixed number

19

RATIONAL VS IRRATIONAL NUMBERS

All numbers are either rational or irrational

*Rational Numbers- Any number that can be expressed as a fraction

Integers

$4 = \frac{4}{1}$

$-3 = \frac{-3}{1}$

$0 = \frac{0}{1}$

Fractions- $\frac{1}{2}$

$-\frac{1}{3}$

Decimals (that terminate)-

$.1 = \frac{1}{10}$

$.25 = \frac{25}{100}$

$-3.2 = -3\frac{2}{10}$

Decimals (that repeat in a pattern)

$.\overline{333} = \frac{1}{3}$

$.\overline{1212} = \frac{4}{33}$

Perfect Squares -

$\sqrt{25} = 5$

$\sqrt{1} = 1$

*Irrational Numbers-

Any decimal or whole number with a decimal that does not repeat in a pattern

Most Popular - π

3.141592654...

(Never repeats in a pattern)

Any radical that is not a perfect square

$\sqrt{15} = 3.87298...$

(never terminates or repeats in a pattern)

PERFECT SQUARES

Perfect Squares

1^2	$1 \cdot 1$	=	1	$\sqrt{1}$	= 1
2^2	$2 \cdot 2$	=	4	$\sqrt{4}$	= 2
3^2	$3 \cdot 3$	=	9	$\sqrt{9}$	= 3
4^2	$4 \cdot 4$	=	16	$\sqrt{16}$	= 4
5^2	$5 \cdot 5$	=	25	$\sqrt{25}$	= 5
6^2	$6 \cdot 6$	=	36	$\sqrt{36}$	= 6
7^2	$7 \cdot 7$	=	49	$\sqrt{49}$	= 7
8^2	$8 \cdot 8$	=	64	$\sqrt{64}$	= 8
9^2	$9 \cdot 9$	=	81	$\sqrt{81}$	= 9
10^2	$10 \cdot 10$	=	100	$\sqrt{100}$	= 10
11^2	$11 \cdot 11$	=	121	$\sqrt{121}$	= 11
12^2	$12 \cdot 12$	=	144	$\sqrt{144}$	= 12
13^2	$13 \cdot 13$	=	169	$\sqrt{169}$	= 13
14^2	$14 \cdot 14$	=	196	$\sqrt{196}$	= 14

ABSOLUTE VALUE

* __Absolute value__ is the distance between 2 numbers + and distance can __not__ be negative

* | | these bars mean absolute value

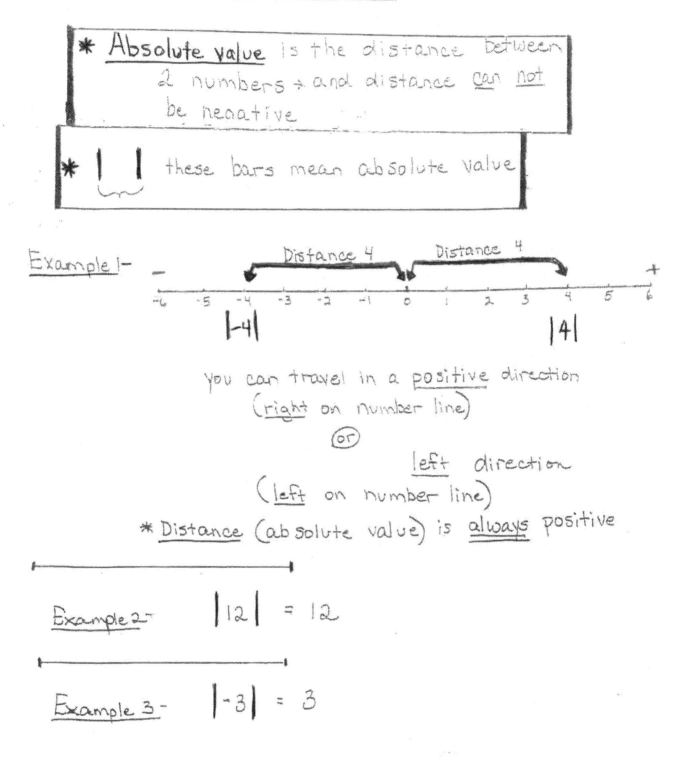

Example 1 -

Distance 4   Distance 4

$|-4|$ $|4|$

you can travel in a __positive__ direction
(__right__ on number line)

(or)

__left__ direction
(__left__ on number line)

* Distance (absolute value) is __always__ positive

Example 2 - $|12| = 12$

Example 3 - $|-3| = 3$

ADDITION AND SUBTRACTION - INTEGERS

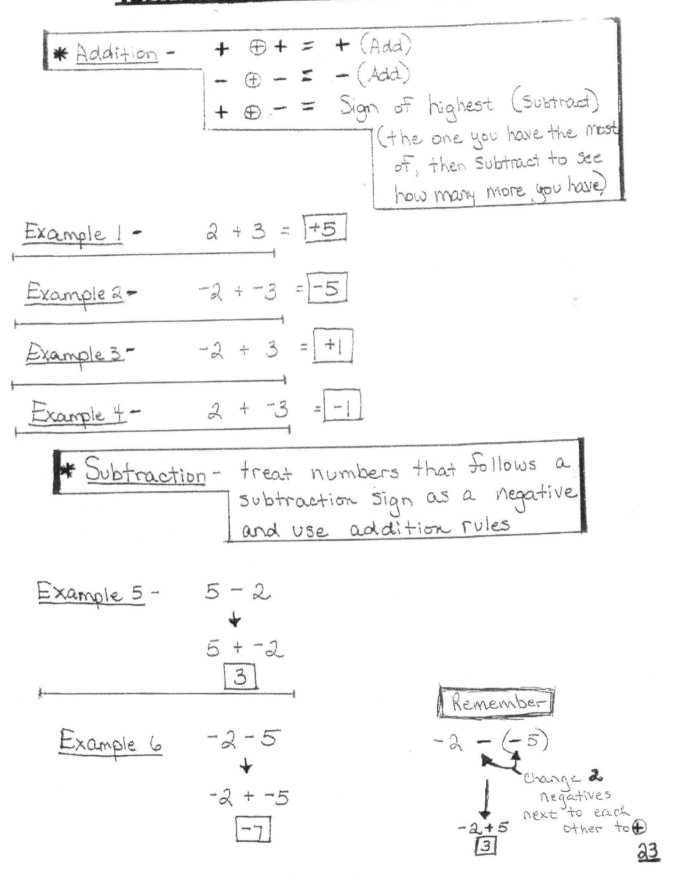

✱ <u>Addition</u> -

  $+ \oplus + = +$ (Add)

  $- \oplus - = -$ (Add)

  $+ \oplus - =$ Sign of highest (Subtract)

  (the one you have the most of, then subtract to see how many more you have)

<u>Example 1</u> - $2 + 3 = \boxed{+5}$

<u>Example 2</u> - $-2 + -3 = \boxed{-5}$

<u>Example 3</u> - $-2 + 3 = \boxed{+1}$

<u>Example 4</u> - $2 + -3 = \boxed{-1}$

✱ <u>Subtraction</u> - treat numbers that follows a subtraction sign as a negative and use addition rules

<u>Example 5</u> - $5 - 2$
                            ↓
                          $5 + -2$
                          $\boxed{3}$

<u>Example 6</u> $-2 - 5$
                            ↓
                          $-2 + -5$
                          $\boxed{-7}$

$\boxed{\text{Remember}}$

$-2 - (-5)$

change 2 negatives next to each other to $\oplus$

$-2 + 5$
$\boxed{3}$

23

MULTIPLICATION AND DIVISION INTEGERS

* **Multiplication** -

$+ \otimes + = +$ (Positive)

$- \otimes - = +$ (Positive)

$+ \otimes - = -$ (Negative)

Example 1 - $2 \times 3 = 6$ $(+ \otimes + = +)$

     or $2(3)$

Example 2 - $(-2) \times (-3) = 6$ $(- \otimes - = +)$

     or $-2(-3)$

Example 3 - $(-2) \times 3 = -6$ $(- \otimes + = -)$

     or $-2(3)$

(IF you multiply 2 negatives, they both become +)

* **Division** -

$+ \div - = -$ (Negative)

$- \div + = -$ (Negative)

$+ \div + = +$ (Positive)

$- \div - = +$ (Positive)

Example 4 - $\dfrac{6}{2} = 3$ (Positive)

     or $6 \div 2$

Example 5 - $\dfrac{-6}{-2} = 3$ (Positive)

     or $-6 \div -2$

Example 6 - $\dfrac{6}{-3} = -2$ (Negative)

     or $6 \div -3$

Example 7 - $\dfrac{-6}{2} = \boxed{-3}$ (Negative)

DISTRIBUTIVE PROPERTY

* Parenthesis must be removed to simplify an expression and solve equations

Step 1. Combine all like terms inside the parenthesis, if possible
2. Multiply each term in parenthesis by number or term outside parenthesis
3. Remove all parenthesis
4. Continue by using order of operations

Example 1 -

$3 + 2(3x + 4)$

$3 + 2(3x + 4)$

$3 + 6x + 8$

$3 + 6x + 8$

Like Terms

$6x + 11$

Example 2 -

$-4(3x + 1 + 5)$

$-4(3x + 1 + 5)$

Like Terms

$-4(3x + 6)$

$-4(3x + 6)$

$-12x - 24$

25

* If there is a $\oplus$ sign right in front of parenthesis, distribute each term by +1 and remove parenthesis

* If there is a $\ominus$ sign right in front of parenthesis, distribute each term by -1 and remove parenthesis

Example 3 -

$$4 + (12x - 5)$$

$$4 + 1(12x - 5) \qquad \underline{Distributive}$$

$$4 + 12x - 5$$

$$\downarrow$$

$$4 + 12x - 5$$

Like Terms

$$\boxed{12x - 1}$$

Example 4 -

$$12x - (4x - 2)$$

$$12x - 1(4x - 2) \qquad \underline{Distributive}$$

$$12x - 4x + 2$$

$$\downarrow$$

$$12x - 4x + 2$$

Like Terms

$$\boxed{8x + 2}$$

26

* If you <u>Can NOT</u> determine Y-intercept

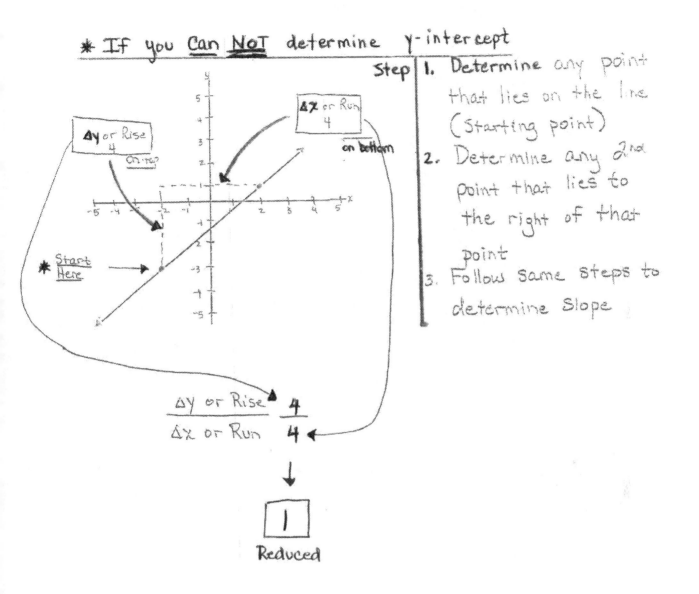

Step 1. Determine any point that lies on the line (Starting point)

2. Determine any 2nd point that lies to the right of that point

3. Follow same steps to determine Slope

Δy or Rise (on top)

Δx or Run 4 (on bottom)

Start Here

$$\frac{\Delta y \text{ or Rise}}{\Delta x \text{ or Run}} \quad \frac{4}{4}$$

↓

$$\boxed{1}$$

Reduced

SPECIAL SLOPES

* Horizontal Lines —   Slope = 0

Example 1-

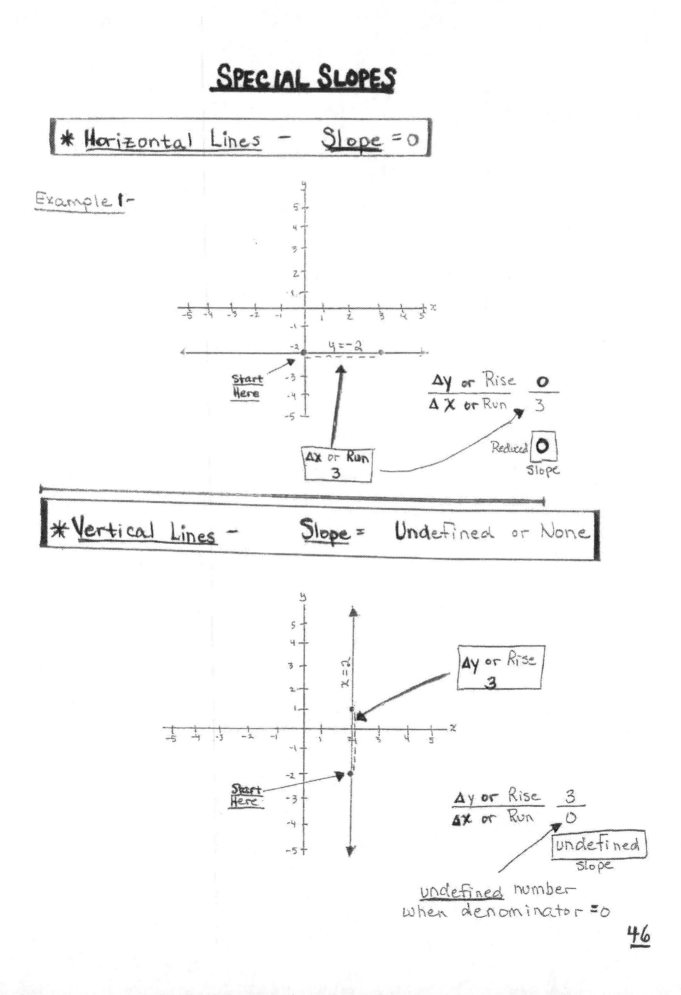

$$\frac{\Delta y \text{ or Rise}}{\Delta x \text{ or Run}} \quad \frac{0}{3}$$

Start Here

Δx or Run
3

Reduced $\boxed{0}$ Slope

* Vertical Lines —   Slope = Undefined or None

Δy or Rise
3

Start Here

$$\frac{\Delta y \text{ or Rise}}{\Delta x \text{ or Run}} \quad \frac{3}{0}$$

undefined Slope

undefined number
when denominator = 0

46

SLOPE DUDE

* Shortcut for remembering slope!

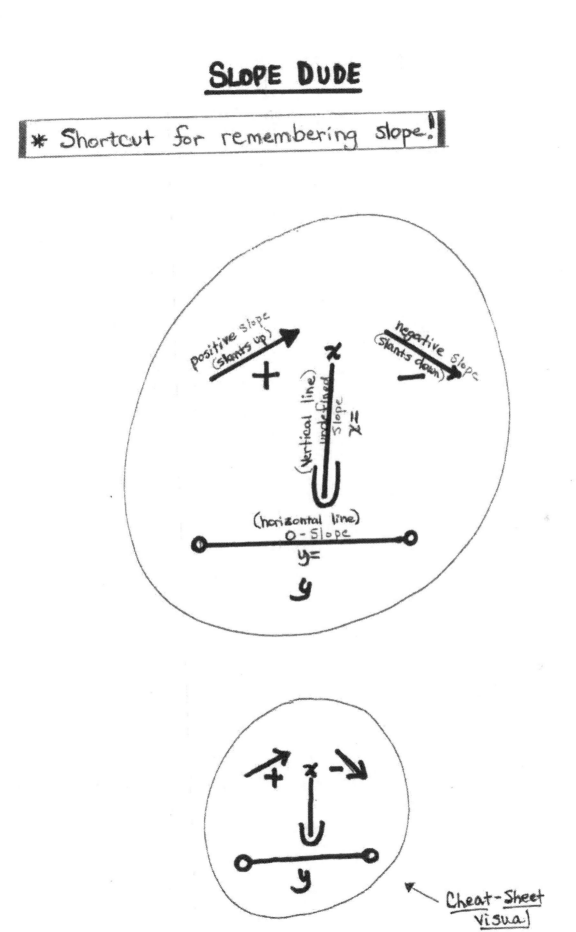

Cheat-Sheet
visual

FINDING SLOPE FROM A TABLE

* Since Slope is a **Rate of Change**, it is **Constant** in all Linear Equations

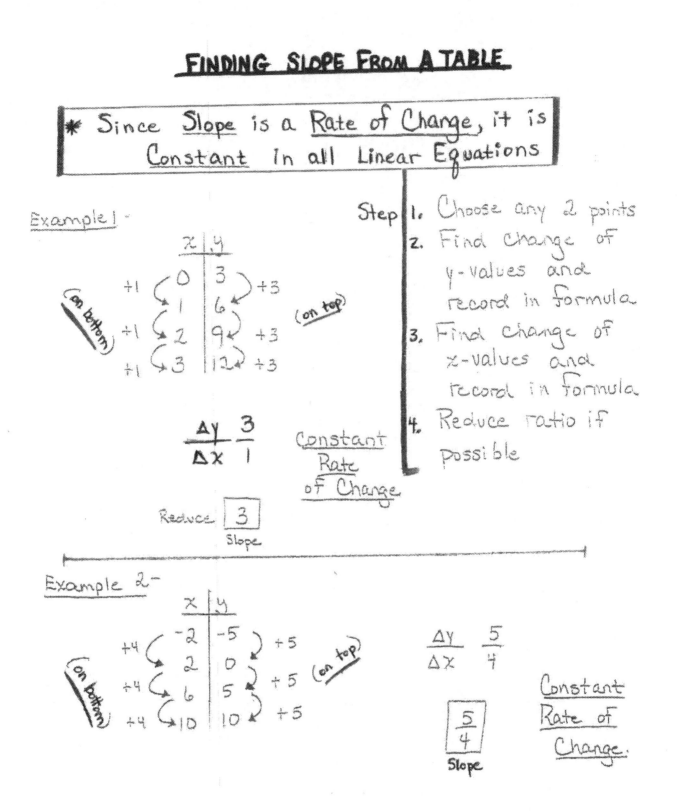

Step 1. Choose any 2 points
2. Find change of y-values and record in formula
3. Find change of x-values and record in formula
4. Reduce ratio if possible

Example 1 -

(on bottom)

x	y
0	3
1	6
2	9
3	12

+1   +3
(on top)

$$\frac{\Delta y}{\Delta x} \quad \frac{3}{1}$$

Constant
Rate
of Change

Reduce $\boxed{3}$
Slope

Example 2 -

(on bottom)

x	y
-2	-5
2	0
6	5
10	10

+4   +5
(on top)

$$\frac{\Delta y}{\Delta x} \quad \frac{5}{4}$$

$\boxed{\frac{5}{4}}$
Slope

Constant
Rate of
Change.

Example 3 -

x	y
8	-8
4	-2
0	4
-4	10

-4 (on bottom)
+6 (on top)

$$\frac{\Delta y}{\Delta x} \quad \frac{6}{-4}$$

Reduce $\boxed{-\dfrac{3}{2}}$

Slope

Constant Rate of Change

Example 4 -

x	y
10	8
6	6
-2	2
-4	1

-4, -8, -2 (on bottom)
-2, -4, -1 (on top)

Reduced

$$\frac{\Delta y}{\Delta x} \quad \frac{-2}{-4} = \frac{1}{2}$$

$$\frac{\Delta y}{\Delta x} \quad \frac{-4}{-8} = \frac{1}{2}$$

$$\frac{\Delta y}{\Delta x} \quad \frac{-1}{-2} = \frac{1}{2}$$

$\boxed{\dfrac{1}{2}}$ Slope

Constant Rates of Change

Example 5 -

x	y
1	4
3	4
5	4
7	4

+2 (on bottom)
0 (on top)

$$\frac{\Delta y}{\Delta x} \quad \frac{0}{2}$$

Reduced $\boxed{0}$ slope

← Horizontal line

Constant Rate of Change

Example 6 -

x	y
2	-4
-2	-4
-6	-4
-10	-4

-1, -1, -1 (on bottom)
-4, -4, -4 (on top)
0, 0, 0

$$\frac{\Delta y}{\Delta x} \quad \frac{-4}{0}$$

$\boxed{\text{undefined slope}}$ ← vertical line

Constant Rate of Change

49

X- and y-INTERCEPTS

* <u>X-intercept</u> is where the graph touches (or intersects) the x-axis

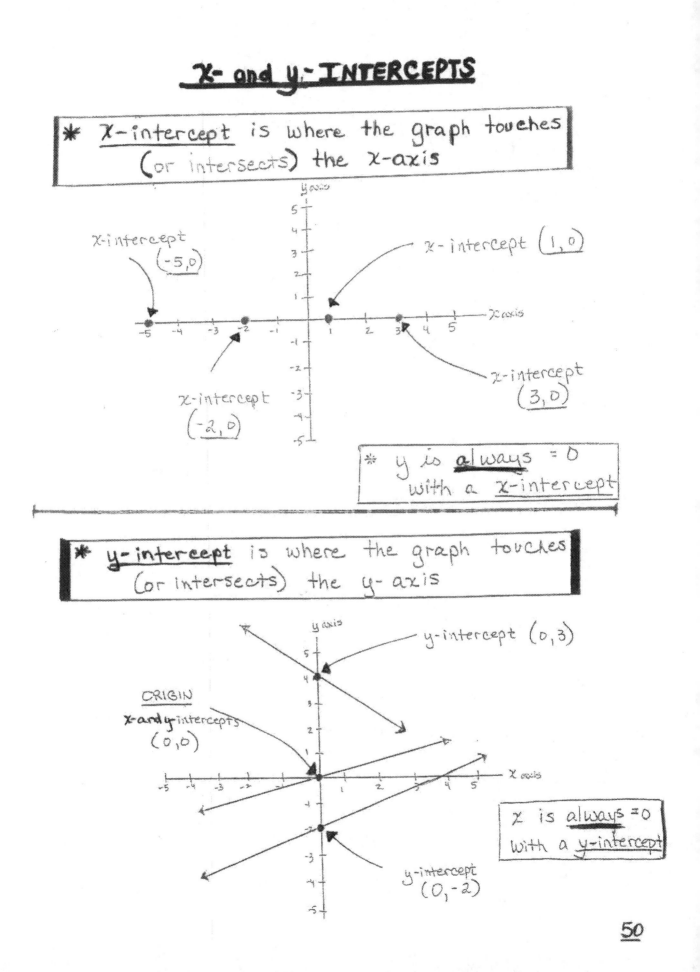

x-intercept (-5,0)

x-intercept (1,0)

x-intercept (-2,0)

x-intercept (3,0)

* y is <u>always</u> = 0 with a <u>x-intercept</u>

* <u>y-intercept</u> is where the graph touches (or intersects) the y-axis

ORIGIN
x-and y-intercepts
(0,0)

y-intercept (0,3)

y-intercept (0,-2)

x is <u>always</u> = 0 with a <u>y-intercept</u>

SLOPE - INTERCEPT FORM

* Slope-intercept form: $y = mx + b$

           Slope   y-intercept
           "Map"   "Begin"

* From this form, you can:
1. Identify Slope
2. Identify y-intercept
3. Graph
4. Use table to make predictions

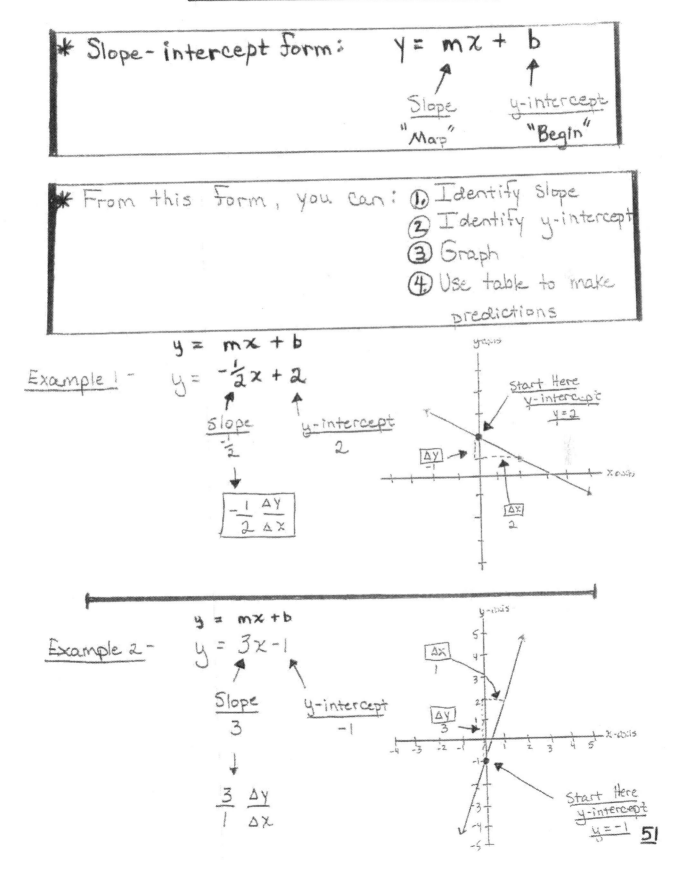

Example 1 -

$y = mx + b$
$y = -\frac{1}{2}x + 2$

slope $-\frac{1}{2}$ y-intercept 2

$$-\frac{1}{2} \quad \frac{\Delta y}{\Delta x}$$

Start Here
y-intercept
y=2

Δy -1
Δx 2

Example 2 -

$y = mx + b$
$y = 3x - 1$

Slope 3   y-intercept -1

$$\frac{3}{1} \quad \frac{\Delta y}{\Delta x}$$

Δx 1

Δy 3

Start Here
y-intercept
y = -1

51

WRITING EQUATIONS FROM TABLES

Step
1. Find rate of change
   (must be constant)
2. Identify y-intercept
   (x must $= 0$)
3. Put information into
   $y = mx + b$

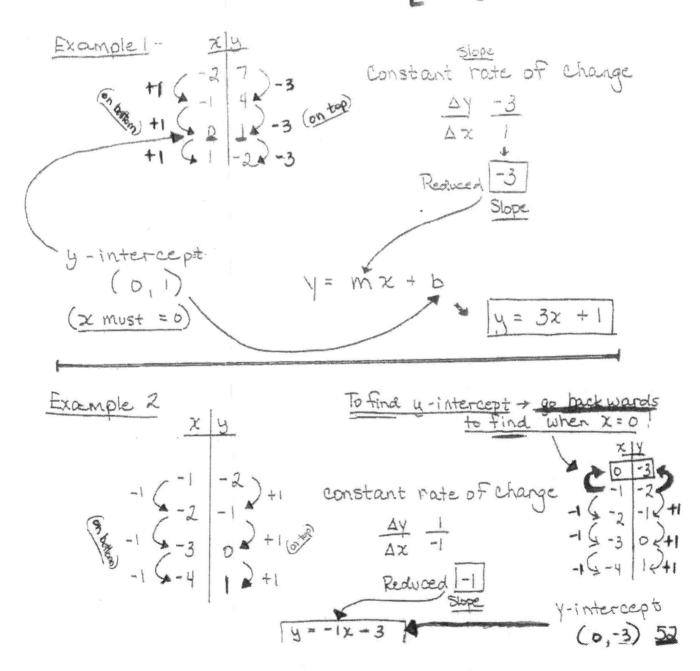

Example 1

x	y
-2	7
-1	4
0	1
1	-2

+1 (on bottom) ... +1 ... +1 ... -3 (on top) ... -3 ... -3

slope
Constant rate of change

$$\frac{\Delta y}{\Delta x} = \frac{-3}{1}$$

Reduced $\boxed{-3}$ Slope

y-intercept
$(0, 1)$
(x must $= 0$)

$y = mx + b$

$\boxed{y = 3x + 1}$

Example 2

x	y
-1	-2
-2	-1
-3	0
-4	1

-1 (on bottom) ... -1 ... -1 ... +1 ... +1 (on top) ... +1

constant rate of change

$$\frac{\Delta y}{\Delta x} = \frac{1}{-1}$$

Reduced $\boxed{-1}$ Slope

$\boxed{y = -1x - 3}$

To find y-intercept → go backwards
to find when $x = 0$!

x	y
0	-3
-1	-2
-2	-1
-3	0
-4	1

-1 ... -1 ... -1 ... +1 ... +1 ... +1

y-intercept
$(0, -3)$ 52

LINEAR EQUATIONS From WORD PROBLEMS

* <u>Slope-intercept form</u> → $y = \underline{m} x + b$

<u>Slope</u>
- per hour
  minute
  pound
  year
- an amount that repeatedly happens over and over

<u>y-intercept</u>
= initial amount
= service fee
= down payment
= starting amount or number
= deposit
= something that happens or is paid one time
= activation fee

<u>Example 1</u> - John goes to the river to tube. To rent a tube, he must pay $20.00 deposit and the rental fee is $3.00 per hour.
Write an equation to express this situation.

$$y = mx + b$$

$$y = 3x + 20$$

or

$$\underset{(cost)}{C} = 3\underset{(hours)}{h} + 20$$

Example 2- Find an equation of a line that passes
through points $\underset{\underset{x_1}{\downarrow}\;\underset{y_1}{\downarrow}}{(1,3)}$ and $(-4,5)$

Step 1. Find slope
     2. Enter either
        point into
        formula

$$\begin{array}{c|c} x & y \\ \hline 1 & 3 \\ -4 & 5 \end{array}$$

(on bottom) -5 $\curvearrowleft$ $+2$ (on top)

$\xleftarrow{\qquad}$ $\dfrac{\text{Find Slope}}{\text{Step 1}}$

$\dfrac{\Delta y}{\Delta x} \quad \dfrac{2}{-5}$

$\boxed{-\dfrac{2}{5}}$

Slope

$(1, 3)$
  $\underset{x_1}{\uparrow} \quad \underset{y_1}{\uparrow}$

$\xleftarrow{\qquad}$ $\dfrac{\text{Choose and label}}{\dfrac{\text{either point}}{\text{Step 2}}}$

$$y - y_1 = m(x - x_1)$$

$$y - 3 = -\frac{2}{5}(x - 1) \quad \longleftarrow \text{Plug into formula}$$

$$y - 3 = -\frac{2}{5}x + \frac{2}{5} \quad \longleftarrow \begin{array}{l}\text{Distributive} \\ \text{Property}\end{array}$$

$+3 \qquad\qquad\qquad +3$

$$\boxed{\begin{array}{l} y = -\frac{2}{5}x + 3\frac{2}{5} \\ y = -\frac{2}{5}x + \frac{17}{5} \end{array}} \quad \longleftarrow \begin{array}{l}\text{Isolate} \\ \text{variable } y\end{array}$$

54

STANDARD FORM TO SLOPE-INTERCEPT

* **Standard form** of a linear equation:
$$\underline{A}x + \underline{B}y = \underline{C}$$

* **How to change standard form to slope-intercept form:**

Step 1. Remove all x-values or constants from y-side of equation by + or − (both sides)

2. Divide y by coeffecient (Itself – on both sides)

Example 1 – $2x - 3y = 12$

$$\cancel{2x} - 3y = 12$$
$$-2x \qquad\qquad -2x$$ } Not like terms

$$-3y = -2x + 12$$

$$\frac{\cancel{-3}y}{-3} = \frac{-2x}{-3} + \frac{12}{-3}$$

$$\boxed{y = \frac{2}{3}x - 4}$$

* **Objective** – to get into $y = mx + b$
why? slope ↗ ↑ y-intercept

Example 2 – $2x + 4y + 3 = 6x - 5$

$$\cancel{2x} + 4y + \cancel{3} = 6x - 5$$
$$-2x \qquad -3 \qquad -2x - 3$$

$$4y = 4x - 8$$

$$\frac{\cancel{4}y}{\cancel{4}} = \frac{4x}{4} - \frac{8}{4}$$

$$\boxed{\begin{array}{l} y = 1x - 2 \\ y = x - 2 \end{array}}$$

PARALLEL AND PERPENDICULAR

* __Parallel__ - when in $y = mx + b$ form (Slope-intercept)
  __Slopes__ __are__ __the__ __Same__

__Example 1__ -
$$y = -\frac{1}{2}x + 1$$
$$y = -\frac{1}{2}x - 4$$

These lines are __parallel__ - __Slopes__ $\left(-\frac{1}{2}\right)$ are the __Same__

** __Perpendicular__ - when in $y = mx + b$ form (Slope-intercept)
  slopes are __BOTH__   - __Opposite signs__
  $\left(\text{one will be } +\right)$
  $\left(\text{one will be } -\right)$

__AND__

- __Slopes__ will be __RECIPROCALS__
(write slope as a fraction and flip the numbers on top and bottom)

__Example 2__ -
$$y = \frac{2}{3}x - 2$$
$$y = -\frac{3}{2}x + 4$$

- These lines are __perpendicular__ -
Slopes are $\frac{2}{3}$ and $-\frac{3}{2}$
(opposite signs __and__ reciprocals)

__Example 3__ -
$$y = 3x - 1$$
$$y = -\frac{1}{3}x + 5$$

*These lines are __perpendicular__ -
Slopes are $\frac{3}{1}$ and $-\frac{1}{3}$
(opposite signs __and__ reciprocals)

GRAPHING LINEAR INEQUALITIES

* **Graphing inequalities** - graph as an equation, shade after graphing.

$$< \; \textcircled{or} \; >$$
↑
Dashed Line

$$\leq \; \textcircled{or} \; \geq$$
↑
Solid Line

Example 1-

Shade Down
↑
$$y \leq 2x - 1$$
↑
Solid Line

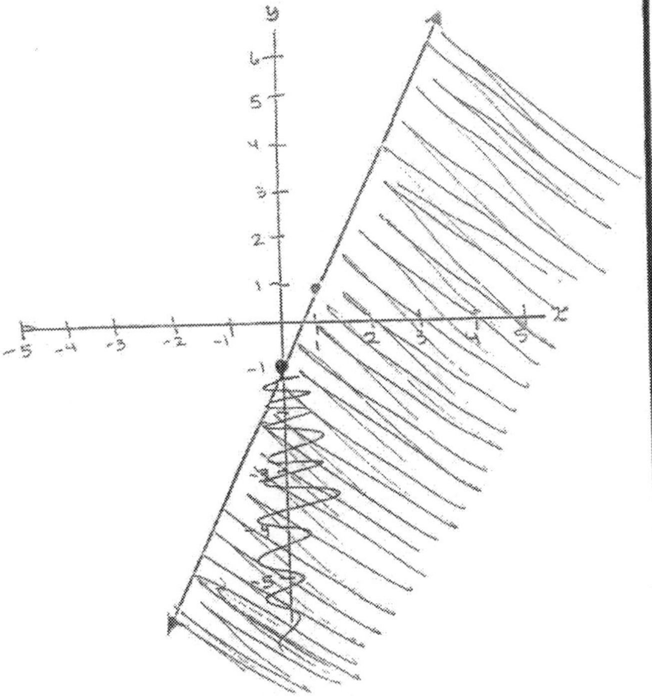

Step 1. Get equation into slope-intercept form $(y = mx + b)$

2. Graph line as usual, but decide if line drawn is <u>dashed</u> or $(< \textcircled{or} >)$ <u>solid</u> $(\leq \text{ or } \geq)$

3. Find where line intersects y-axis
 - If $< \textcircled{or} \leq$, shade <u>down</u> y-axis and remainder of graph below line
 - If $> \textcircled{or} \geq$, shade <u>up</u> y-axis and remainder of graph above line

* When <u>dividing</u> or <u>multiplying</u> an **inequality** by a <u>Negative</u> number to rewrite into slope-intercept form $(y = mx + b)$,

MUST DO!

<u>Always</u> <u><u>reverse</u></u> inequality sign.

Example 2 -

$$2x - 3y < 6$$

$$\cancel{2x} - 3y < 6 \quad \rbrace \text{ Not like Terms}$$
$$\underline{-2x \qquad\qquad -2x}$$
$$-3y < -2x + 6$$

If you $\div$
by a Negative
↓
<u>MUST Reverse</u>
<u>INEQUALITY</u>
<u>Sign</u>

$$\frac{\cancel{-3}y}{\cancel{-3}} < \frac{-2x}{-3} + \frac{6}{-3}$$

$$y > \frac{2}{3}x - 2$$

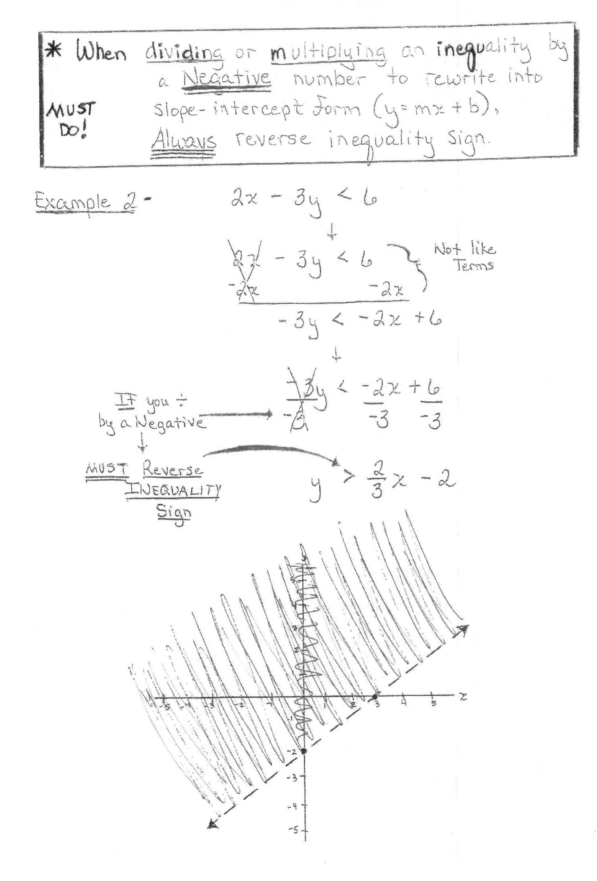

FUNCTIONS

* A Function is a relation, (x, y), in which **all** x-values are different/unique

* Ordered Pairs and Tables - x-values can not repeat

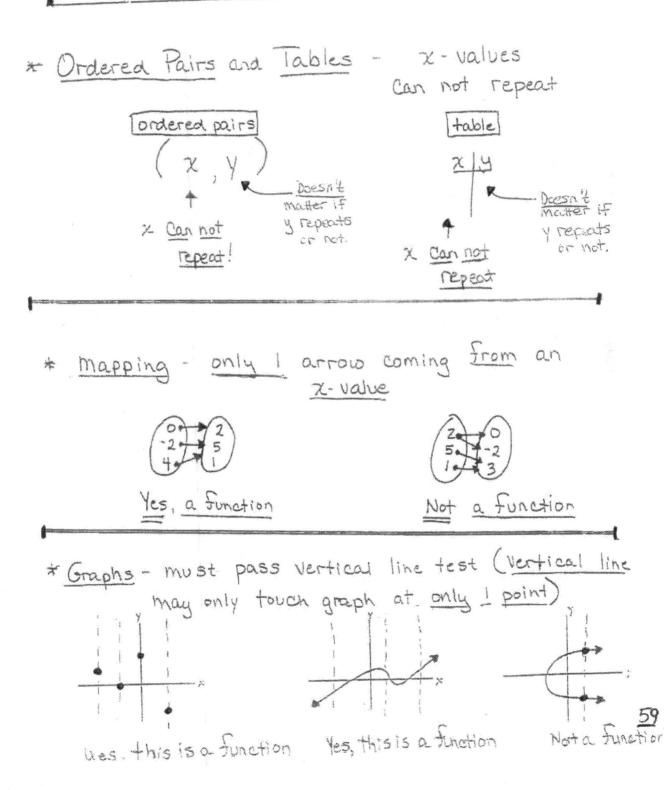

ordered pairs

(x , y)

x Can not repeat!

Doesn't matter if y repeats or not.

table

$x | y$

x Can not repeat

Doesn't matter if y repeats or not.

* mapping - only 1 arrow coming from an x-value

Yes, a function

Not a function

* Graphs - must pass vertical line test (vertical line may only touch graph at only 1 point)

Yes, this is a function

Yes, this is a function

Not a function

59

DOMAIN AND RANGE

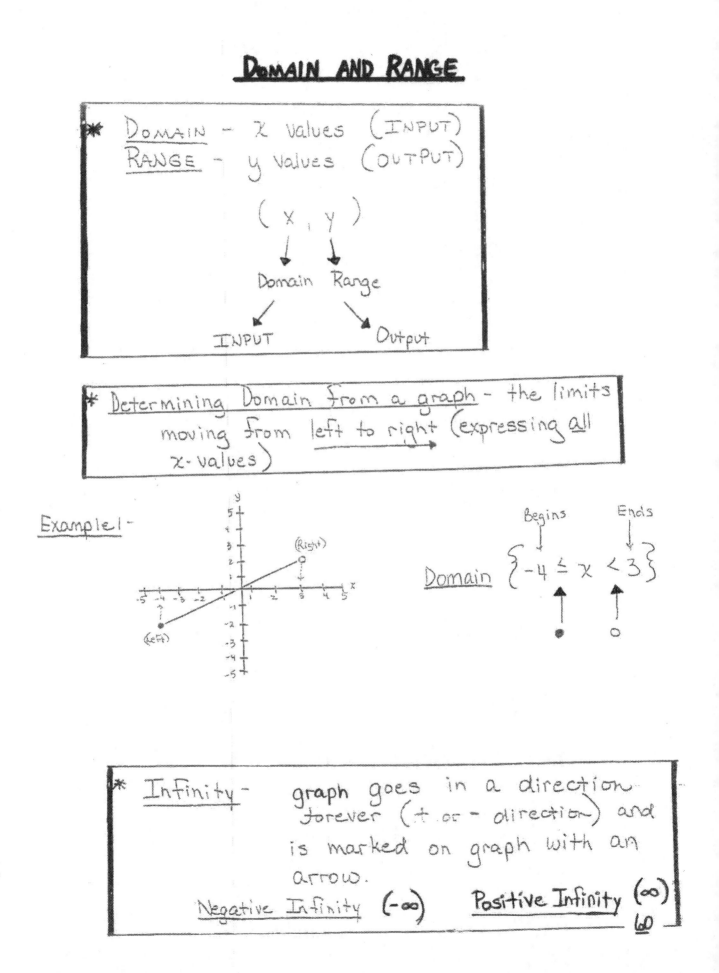

Domain - x values (INPUT)
Range - y values (OUTPUT)

(x , y)

Domain   Range

INPUT   Output

---

* Determining Domain from a graph - the limits moving from left to right (expressing all x-values)

Example 1-

Domain $\{-4 \leq x < 3\}$

Begins   Ends

---

* Infinity - graph goes in a direction forever (+ or - direction) and is marked on graph with an arrow.

Negative Infinity ($-\infty$)   Positive Infinity (∞)

60

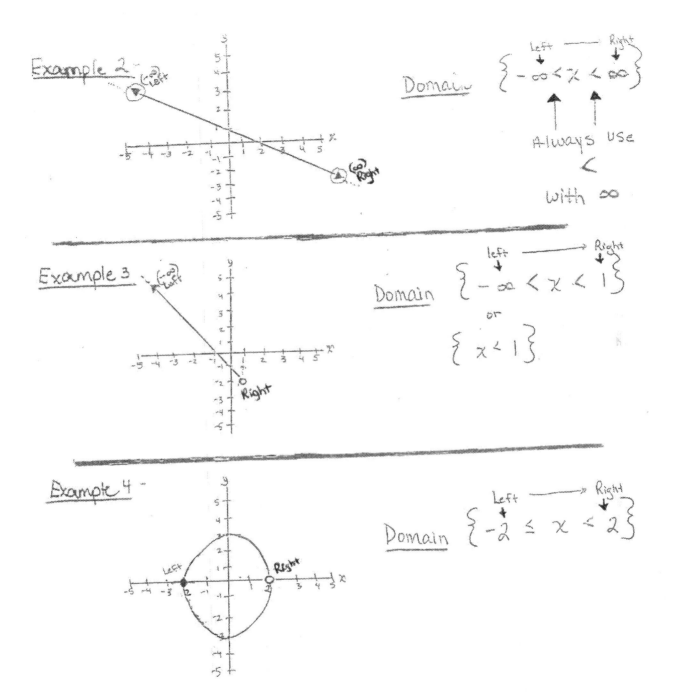

Example 2

Domain $\{ -\infty < x < \infty \}$

Left ⟶ Right

Always use < with ∞

Example 3

Domain $\{ -\infty < x < 1 \}$

or

$\{ x < 1 \}$

Left ⟶ Right

Example 4

Domain $\{ -2 \le x < 2 \}$

Left ⟶ Right

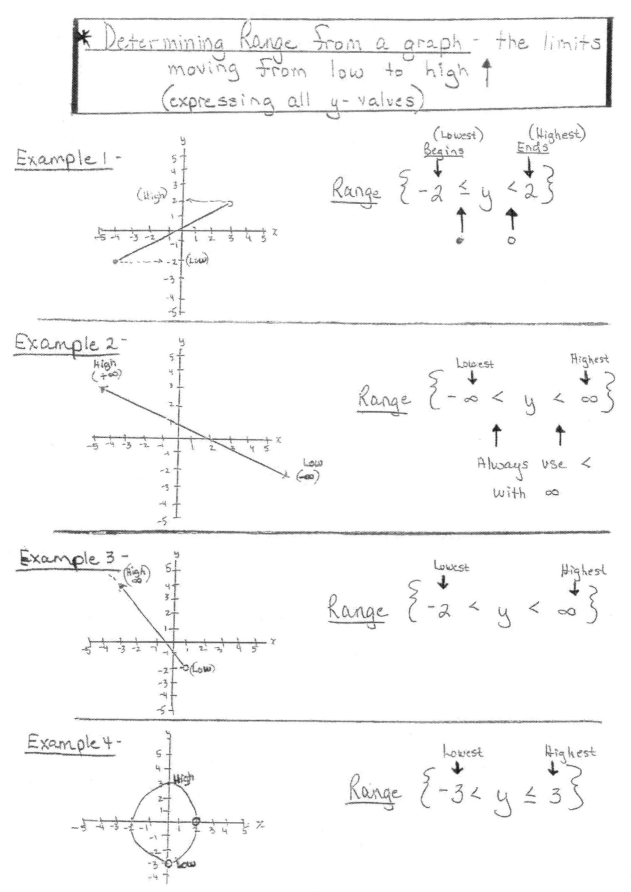

* Determining Range from a graph - the limits moving from low to high ↑ (expressing all y-values)

Example 1 -

(High)

(Low)

$$\text{Range} \left\{ -2 \leq y < 2 \right\}$$

(Lowest) Begins   (Highest) Ends

Example 2 -

High (+∞)

Low (-∞)

$$\text{Range} \left\{ -\infty < y < \infty \right\}$$

Lowest   Highest

Always use < with ∞

Example 3 -

(High ∞)

(Low)

$$\text{Range} \left\{ -2 < y < \infty \right\}$$

Lowest   Highest

Example 4 -

High

Low

$$\text{Range} \left\{ -3 < y \leq 3 \right\}$$

Lowest   Highest

Solving Systems by Graphing

* **System** – The intersection of two or more graphs; equations that share the same solutions

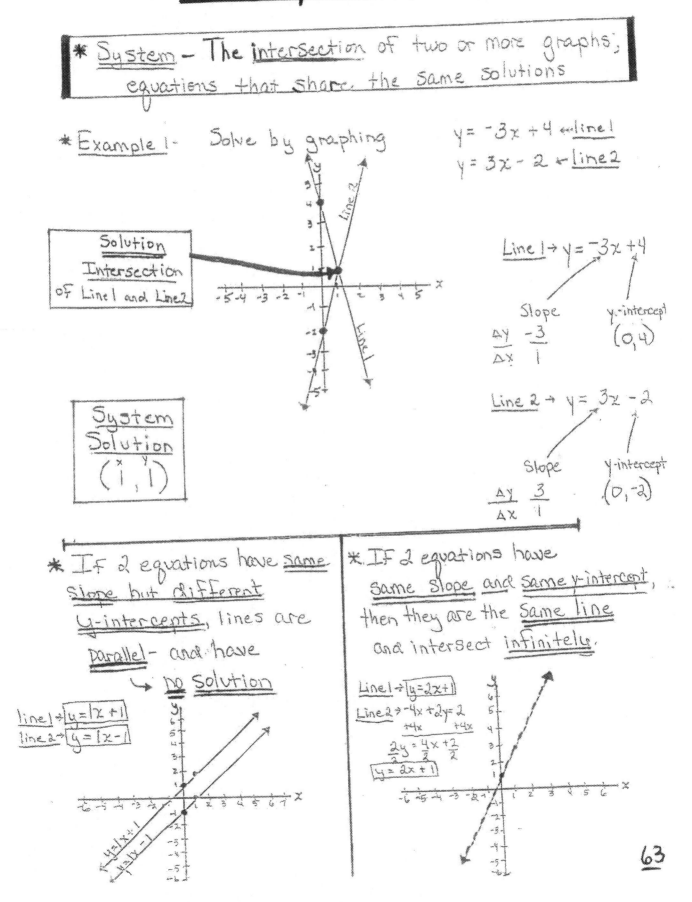

* **Example 1** - Solve by graphing

$$y = -3x + 4 \leftarrow line1$$
$$y = 3x - 2 \leftarrow line2$$

Solution Intersection of Line1 and Line2

System Solution
$$\binom{x}{1}, \binom{y}{1}$$

Line1 → $y = -3x + 4$

Slope
$$\frac{\Delta y}{\Delta x} \quad \frac{-3}{1}$$

y-intercept
$(0, 4)$

Line2 → $y = 3x - 2$

Slope
$$\frac{\Delta y}{\Delta x} \quad \frac{3}{1}$$

y-intercept
$(0, -2)$

* If 2 equations have **same** **slope** but **different** **y-intercepts**, lines are **Parallel** - and have → **NO Solution**

line1 → $y = 1x + 1$
line2 → $y = 1x - 1$

* If 2 equations have **same slope** and **same y-intercept**, then they are the **same line** and intersect **infinitely**.

Line1 → $y = 2x + 1$
Line2 → $-4x + 2y = 2$
$$\quad +4x \quad\quad +4x$$
$$\frac{2y}{2} = \frac{4x + 2}{2}$$
$$y = 2x + 1$$

63

SOLVING SYSTEMS BY SUBSTITUTION

Step 1. Find one equation written in $x=$ or $y=$

or

rewrite one equation into $x=$ or $y=$

2. Substitute the $x=$ or $y=$ into the other equation for the x or y variable (substitute with parenthesis) and solve for variable

3. Substitute back into original equation to solve for other variable

Example 1 -
$$y = 3x - 2 \longrightarrow$$ This is in $y=$
$$x - y = 4$$

So
substitute $(3x-2)$ from $y=(3x-2)$ into the other equation for y and solve for x

$$x - (\overset{y}{3x} - 2) = 4$$

$$x - 1(3x - 2) = 4$$

$$x - 3x + 2 = 4$$

$$-2x + 2 = 4$$
$$ -2 \quad -2$$

$$\frac{-2x}{-2} = \frac{2}{-2}$$

$$\boxed{x = -1}$$

— because $x = -1$, Substitute -1 for x
— because systems have the same solution

Step 3
$$y = 3x - 2$$

$$y = 3(\overset{x}{-1}) - 2$$

$$y = -3 - 2$$

$$\boxed{y = -5}$$

Solution
$$\boxed{(-1, -5)}$$

64

SOLVING SYSTEMS BY ELIMINATION (BY ADDITION)

Step 1. Line up a x- and y-values, **equal** signs and constants

2. Eliminate x or y- values (doesn't matter which one - pick easiest) by adding to get 0 and add remaing variable to solve (May have to multiply either row 1 or 2 (or both) to eliminate)

3. Substitute this found value into either original equation to solve for other variable

Example 1:
$$2x + 3y = 14$$
$$-2x + 7y = 16$$

$\downarrow$

Eliminate x-values by addition

$$\cancel{2x} + 3y = 14$$
$$\cancel{-2x} + 7y = 16$$
$$\overline{\quad 10y = 30}$$

$\downarrow$

$$\frac{10y}{10} = \frac{30}{10}$$

$$\boxed{y = 3}$$

because systems have the same solution

Substitute $y=3$ into either original equation and solve for x

$$2x + 3y = 14$$
$$2x + 3(3) = 14$$
$$2x + 9 = 14$$
$$\quad -9 \quad -9$$
$$\overline{\frac{2x}{2} = \frac{5}{2}}$$

$$\boxed{x = \frac{5}{2}}$$

Solution
$$\left(\frac{5}{2}, 3\right)$$
$$x \quad y$$
$$(2.5, 3)$$

65

Example 2-

$$4x + 5y = 19 \quad \leftarrow \text{Row 1}$$
$$2x - 4y = -10 \quad \leftarrow \text{Row 2}$$

Eliminate x-values by multiplying row 2 by -2

(Sometimes x and y values do not eliminate by addition; So you must manipulate one of them to make them eliminate! Doesn't matter which one you choose to manipulate to eliminate.)

$$4x + 5y = 19 \quad \leftarrow \text{Row 1}$$
$$-2(2x - 4y = -10) \quad \leftarrow \text{Row 2}$$

$$\downarrow$$

$$4x + 5y = 19 \quad \leftarrow \text{Row 1}$$
$$-4x + 8y = 20 \quad \leftarrow \underline{\text{New Row 2}}$$
$$\overline{\qquad 13y = 39}$$

$$\downarrow$$

$$\frac{13y}{13} = \frac{39}{13}$$

$$\boxed{y = 3}$$

Substitute y=3 into either original equation and solve for x

$$4x + 5y = 19$$
$$4x + 5(3) = 19$$
$$4x + 15 = 19$$
$$\quad -15 \quad -15$$
$$\overline{\qquad\qquad}$$
$$\frac{4x}{4} = \frac{4}{4}$$

$$\boxed{x = 1}$$

Solution

$$\left(\underset{x}{1} , \underset{y}{3} \right)$$

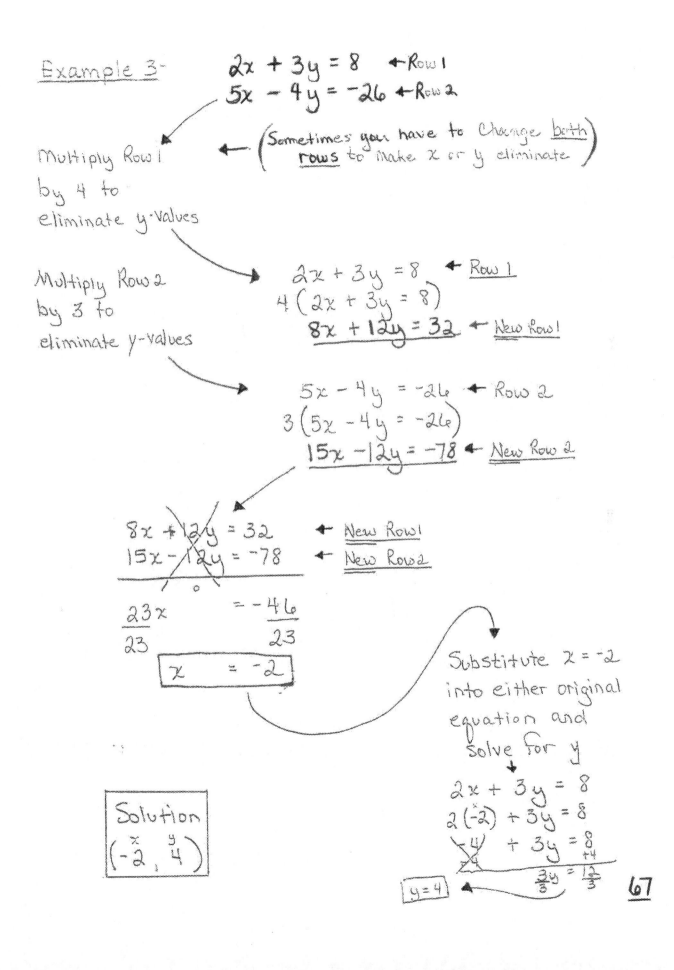

Example 3:

$$2x + 3y = 8 \quad \leftarrow \text{Row 1}$$
$$5x - 4y = -26 \quad \leftarrow \text{Row 2}$$

$\leftarrow$ (Sometimes you have to Change **both** rows to make x or y eliminate)

Multiply Row 1 by 4 to eliminate y-values

Multiply Row 2 by 3 to eliminate y-values

$$2x + 3y = 8 \quad \leftarrow \underline{\text{Row 1}}$$
$$4(2x + 3y = 8)$$
$$\underline{8x + 12y = 32} \quad \leftarrow \underline{\text{New Row 1}}$$

$$5x - 4y = -26 \quad \leftarrow \text{Row 2}$$
$$3(5x - 4y = -26)$$
$$\underline{15x - 12y = -78} \quad \leftarrow \underline{\text{New Row 2}}$$

$$8x + 12y = 32 \quad \leftarrow \underline{\text{New Row 1}}$$
$$15x - 12y = -78 \quad \leftarrow \underline{\text{New Row 2}}$$

$$\frac{23x}{23} = \frac{-46}{23}$$

$$\boxed{x = -2}$$

Substitute $x = -2$ into either original equation and solve for y

$$2x + 3y = 8$$
$$2(-2) + 3y = 8$$
$$-4 + 3y = 8$$
$$\frac{+4 \quad +4}{\frac{3y}{3} = \frac{12}{3}}$$
$$\boxed{y = 4}$$

Solution
$$\left(\overset{x}{-2}, \overset{y}{4} \right)$$

67

EXPONENT RULES

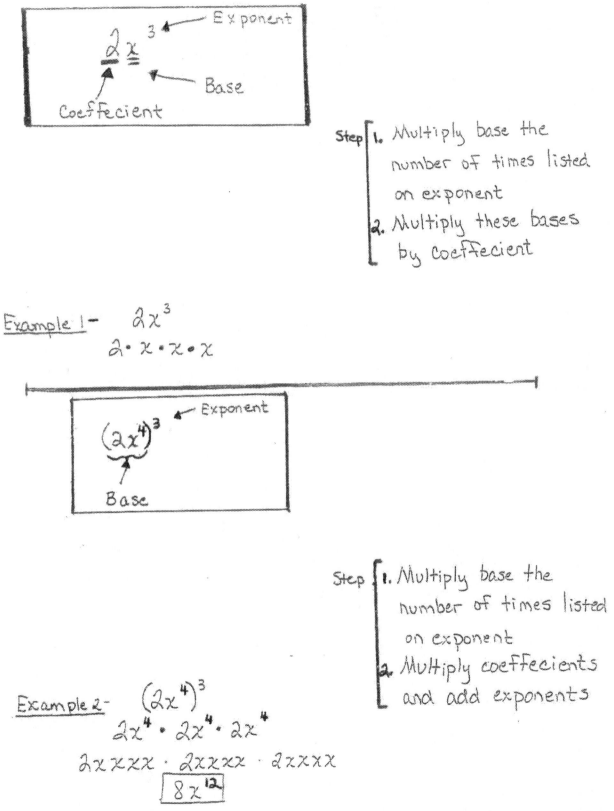

$$2\underline{\underline{x}}^{3}$$

← Exponent

← Base

Coeffecient

Step
1. Multiply base the number of times listed on exponent
2. Multiply these bases by coeffecient

Example 1 - $2x^3$

$2 \cdot x \cdot x \cdot x$

$(2x^4)^3$

← Exponent

Base

Step
1. Multiply base the number of times listed on exponent
2. Multiply coeffecients and add exponents

Example 2 - $(2x^4)^3$

$2x^4 \cdot 2x^4 \cdot 2x^4$

$2xxxx \cdot 2xxxx \cdot 2xxxx$

$\boxed{8x^{12}}$

MULTIPLYING EXPONENTS

* Bases **must** be the same to multiply variables with exponents

Step
1. Bases **must** be the same
2. If bases are the same, add all exponents for each base.

Example 1 - $x^4 \cdot x^3$

$(4+3)$

x^7

* If bases have a coeffecient, multiply coeffecients, then add like bases' exponents

Step
1. Bases **must** be the same
2. Multiply coeffecients, then add exponents for like bases

Example 2 - $4x^4 \cdot 2x^3$

$(4 \cdot 2)(x^{4+3})$

$8x^7$

* If more than 1 base, add all exponents for each base
* If no exponent, there is an understood "1" as an exponent

Example 3 - $6a^2b \cdot 3a^3b^6c^4$

$(6 \cdot 3)(a^{2+3})(b^{1+6})(c^4)$

$18 \quad a^5 \quad b^7 \quad c^4$

$18a^5b^7c^4$

69

DIVIDING EXPONENTS

* Bases __must__ be the same **to divide variables with** exponents

Step 1. Bases __must__ be the same to divide
2. If the bases are the same, __subtract__ exponents

Example 1- $\dfrac{x^6}{x^4}$

$(6-4)$

$\dfrac{x^2}{1}$ or x^2

* If bases have coeffecients, reduce coeffecients (same as reducing a fraction) first, then subtract coeffecients.

Step 1. Bases __must__ be the same to divide
2. Reduce coeffecients, subtract exponents

Example 2- $\dfrac{16x^5}{12x^7}$

Reduce $\dfrac{16 \div 4}{12 \div 4} = \dfrac{4}{3}$

$\dfrac{4x^5}{3x^7}$

$(5-7)$
-2

$\boxed{\dfrac{4}{3}x^{-2} \ \text{or} \ \dfrac{4}{3x^2}}$

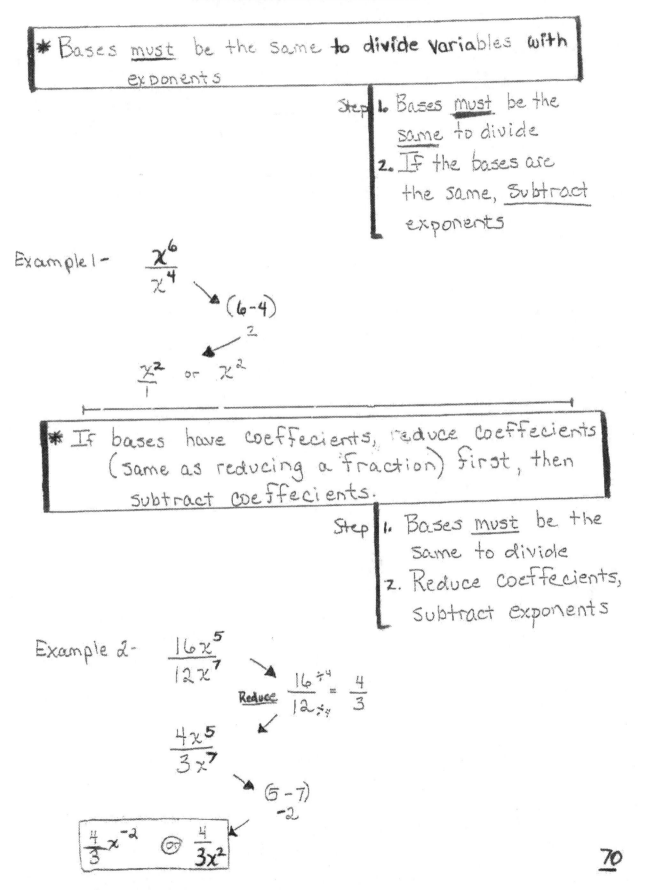

ADDING AND SUBTRACTING EXPONENTS

* __All__ bases and their exponents must match exactly to add or subtract terms with exponents. Coeffecients __do NOT__

Step 1. IF all bases and their exponents match __exactly__, add or subtract coeffecients

2. __Do NOT__ change exponents

Example 1. $2x^3 + 5x^3$

2 + 5
7

← variables and exponents match exactly

$\boxed{7x^3}$

Example 2. $3x^2y^3 - 8x^2y^3$

← variables and exponents match exactly

3 - 8
-5

$\boxed{-5x^2y^3}$

Example 3. $4x^2 + 2x$

← variables and exponents __do not__ match exactly

Can not simplify (exponents do __not__ match)

Example 4. $3x^2y + 6xy^2$

← variables and exponents __do not__ match exactly

Can not simplify (Bases and exponents do __not__ match)

71

NEGATIVE EXPONENTS

> * If exponents are negative, you can move that base with its exponent from the top of a fraction bar to the bottom (or) **from** the bottom of a fraction bar to the top by switching the − sign to a positive

Example 1- $\dfrac{a^{-2}}{1} = \boxed{\dfrac{1}{a^2}}$

Example 2- $\dfrac{1}{b^{-5}} = \boxed{\dfrac{b^5}{1}}$

Example 3- $\dfrac{x^3}{1} = \boxed{\dfrac{1}{x^{-3}}}$

Example 4- $\dfrac{1}{y^4} = \boxed{\dfrac{y^{-4}}{1}}$

> * To simplify exponents, it is <u>easier</u> to move all negative exponents with its base to the opposite side of the fraction bar <u>before</u> simplifying

Example 5-

$$\dfrac{x^{-3}\,y^4}{x^5\,y^{-2}}$$

$$\dfrac{x^{-3}\;y^2}{x^3\;y^{-2}}$$

$$\dfrac{y^4 \cdot y^2}{x^3 \cdot x^5} \longrightarrow \boxed{\dfrac{y^6}{x^4}}$$

72

EXPONENTIAL GROWTH and DECAY

Equation → $y = a(b)^x$ ← Time = Minutes
$=$ Days
$=$ Months
$-$ Years

a → Initial Amount
or
Beginning Number
Starting Amount

b → Increase-Growth
or
Decrease-Decay

IF $b > 1$ (Growth)
IF $b < 1$ (Decay)

Exponential Growth
Graph →

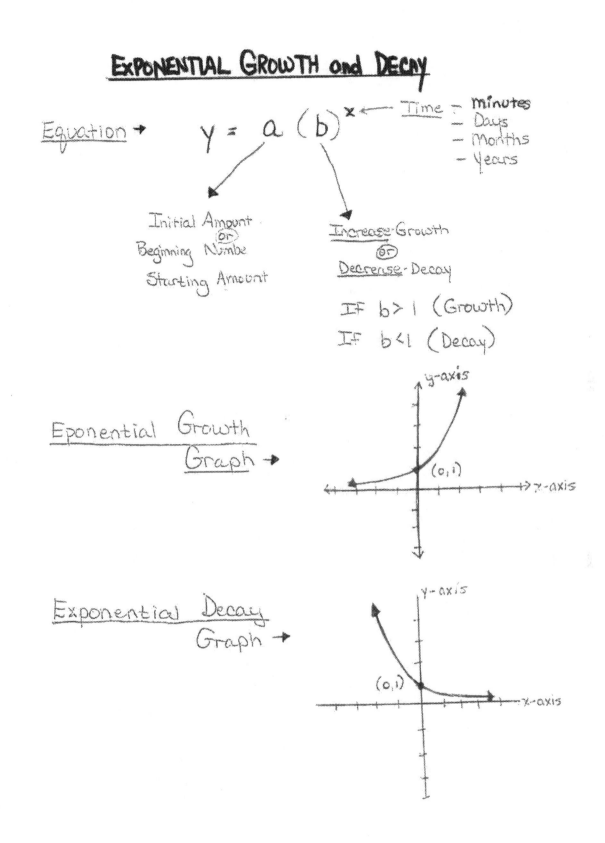

Exponential Decay
Graph →

73

SOLVING EXPONENTIAL EQUATIONS

> **IF** bases are the same, set exponents equal and solve.

Example 1 - $\quad 3^8 = 3^{x+2}$

$\downarrow$

Bases are the same
so set exponents equal

$\downarrow$

$$8 = x + 2$$
$$-2 \qquad\qquad -2$$

$\downarrow$

$$\boxed{6 = x}$$

> **IF** bases are <u>not</u> the same, rewrite base(s) so that they are all the same.

Example 2 - $\quad 9^{x+3} = \dfrac{1}{3}$

$\downarrow$

Rewrite 9 as 3^2 $\quad\longrightarrow\quad 9^{x+3} = 3^{-1} \quad$ Rewrite $\dfrac{1}{3^1}$ to 3^{-1}

$\downarrow$

$$3^{2(x+3)} = 3^{-1}$$

Bases are the same
so set exponents equal

$\downarrow$

$$2(x+3) = -1$$
$$2x + 6 = -1$$
$$\quad -6 \quad\quad -6$$

$\longrightarrow \quad \dfrac{2x}{2} = \dfrac{-7}{2}$

$$\boxed{x = \dfrac{-7}{2}}$$

74

Prime Numbers and Prime Factoring

* **Prime numbers** are numbers that only have 2 factors. (2 different factors)

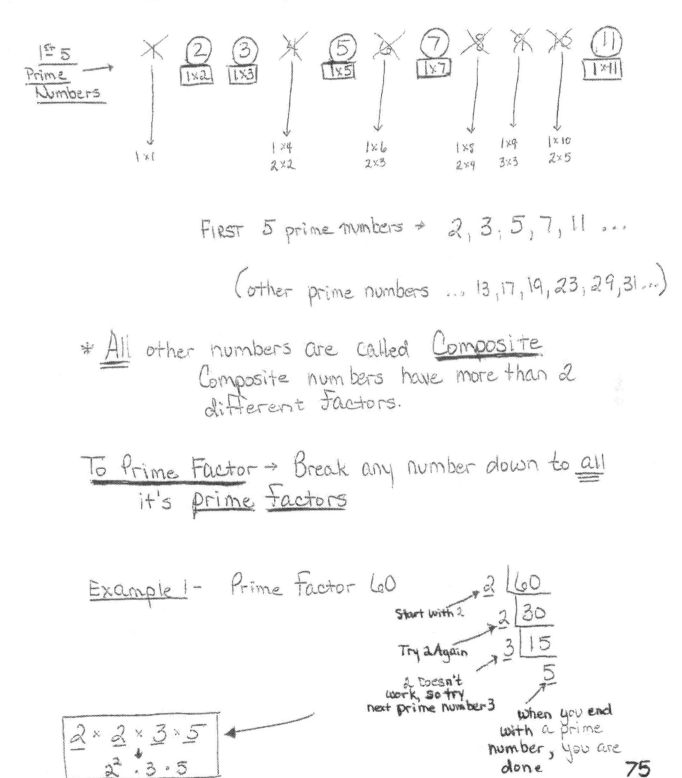

First 5 prime numbers → 2, 3, 5, 7, 11 ...

(other prime numbers ... 13, 17, 19, 23, 29, 31 ...)

* **All** other numbers are called <u>Composite</u>. Composite numbers have more than 2 different factors.

<u>To Prime Factor</u> → Break any number down to <u>all</u> it's <u>prime factors</u>

Example 1 - Prime Factor 60

Start with 2

Try 2 Again

2 Doesn't work, so try next prime number 3

When you end with a prime number, you are done

$2 \times 2 \times 3 \times 5$
$2^2 \cdot 3 \cdot 5$

FACTORS vs PRIME FACTORS

Factors – <u>Any</u> number that divides <u>evenly</u> into another number

Prime Factors – <u>Any</u> prime number that divides <u>evenly</u> into another number (2, 3, 5, 7, 11, ...)

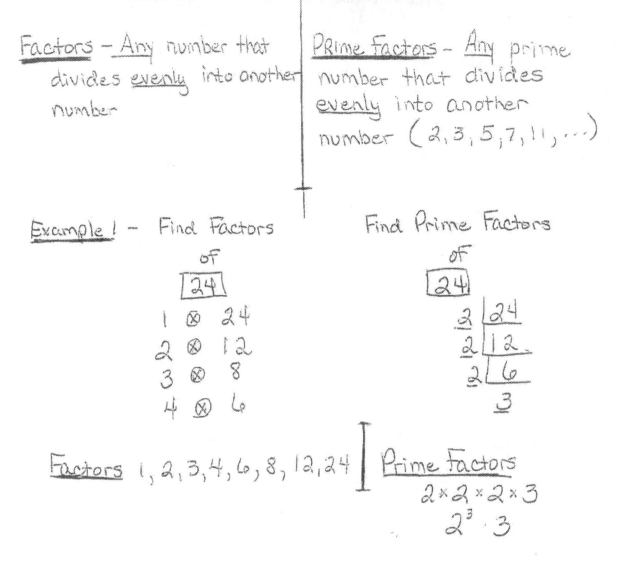

Example 1 – Find Factors of

$\boxed{24}$

1 ⊗ 24
2 ⊗ 12
3 ⊗ 8
4 ⊗ 6

Find Prime Factors of

$\boxed{24}$

2 | 24
2 | 12
2 | 6
    3

<u>Factors</u> 1, 2, 3, 4, 6, 8, 12, 24

<u>Prime Factors</u>
$2 \times 2 \times 2 \times 3$
$2^3 \cdot 3$

SIMPLIFYING RADICALS

* **Prime Numbers**- Numbers that only have 2 factors, 1 and itself

First seven prime numbers- 2, 3, 5, 7, 11, 13, 17...

* **Prime Factoring**- Dividing a number down to only prime numbers

Step 1. Take number out of radical
2. Divide number by prime numbers to find all it's prime factors
3. Put all prime factors back under radical
4. Any pair of prime factors is a perfect square. Bring out all perfect squares. (If more than one square root is brought out, find their product by multiplying.

Example 1- $\sqrt{20}$

```
        20
       /  \
    (2)    10
   Prime  /  \
       (2)   (5)
      Prime  prime
```

$\sqrt{2 \times 2} \; 5$
perfect
square
"4"

$$2\sqrt{5}$$

Example 2 - $\sqrt{48}$ →

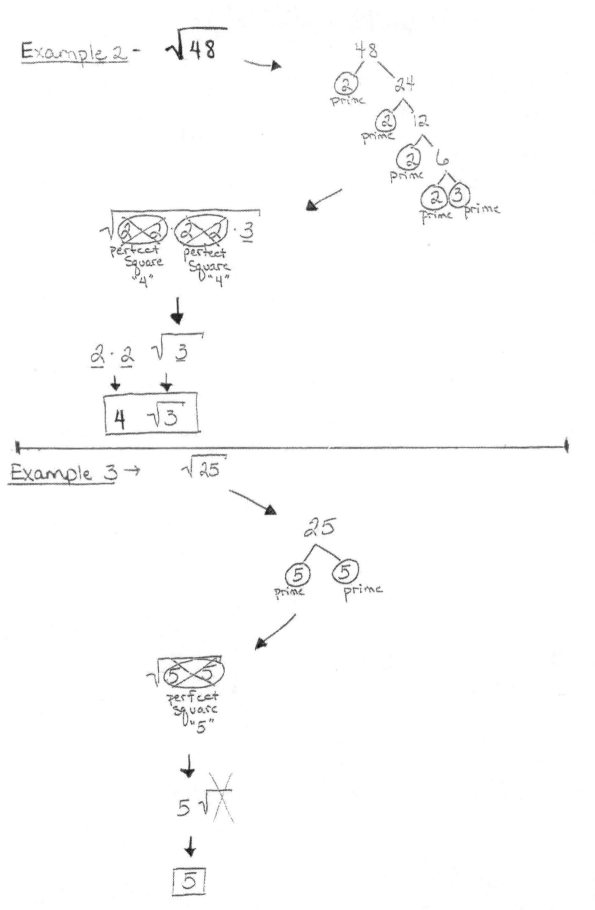

$$2 \cdot 2 \, \sqrt{3}$$

$$\boxed{4 \quad \sqrt{3}}$$

Example 3 → $\sqrt{25}$

$$5 \, \sqrt{\cancel{}}$$

$$\boxed{5}$$

RADICALS WITH VARIABLES

* If variables have an <u>even</u> exponent, bring out the variable with half the exponent
* If variables have an <u>odd</u> exponent, bring out the variable with half the exponent and odd remaining variable will be left inside radical

Step 1. Even exponents, pair up variables and bring out pairs. <u>No</u> remaining variables

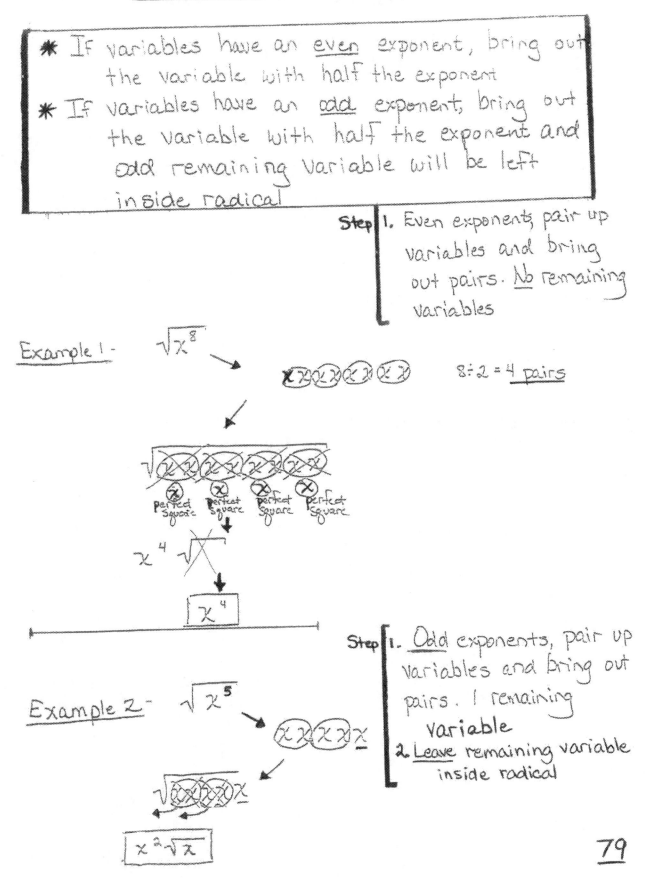

Example 1 - $\sqrt{x^8}$

$8 \div 2 = 4$ pairs

Perfect Square   Perfect Square   Perfect Square   Perfect Square

$x^4 \sqrt{}$

x^4

Step 1. <u>Odd</u> exponents, pair up variables and bring out pairs. 1 remaining variable
2. <u>Leave</u> remaining variable inside radical

Example 2 - $\sqrt{x^5}$

$x^2 \sqrt{x}$

ADDING AND SUBTRACTING RADICALS

* Radicals **must** be the <u>same</u> to <u>add</u> or <u>subtract</u>. If radicals are the <u>same</u>, add or <u>subtract</u> <u>coeffecients</u>. Radicals do <u>not</u> change.

Step 1. Radicals must be the same

2. If radicals are the same, add or subtract coeffecients.

3. <u>Do</u> <u>not</u> change radical

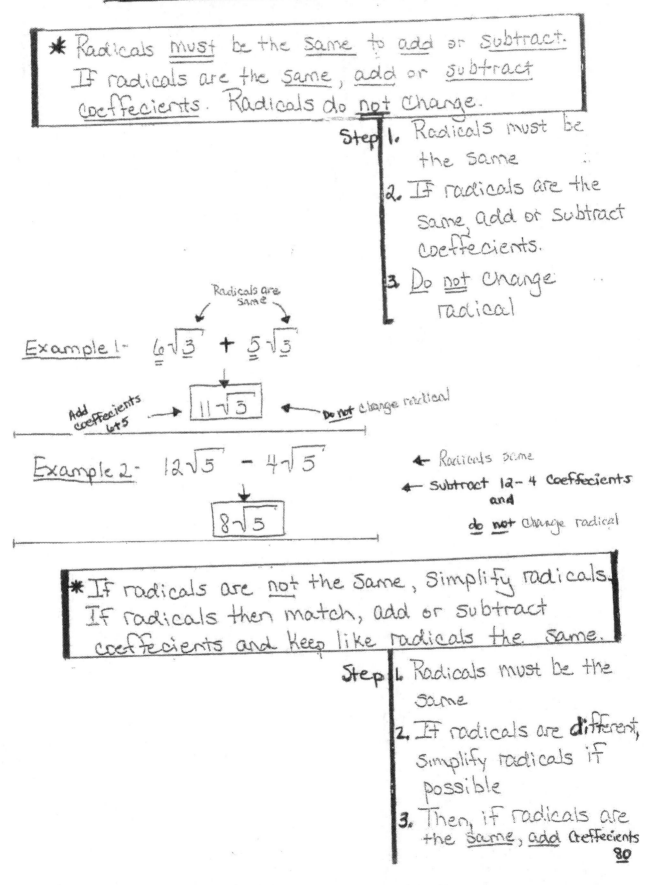

Radicals are same

Example 1- $6\sqrt{3}$ + $5\sqrt{3}$

Add coeffecients 6+5 → $11\sqrt{5}$ ← Do not change radical

Example 2- $12\sqrt{5}$ - $4\sqrt{5}$ ← Radicals same

← Subtract 12-4 Coeffecients and

<u>do</u> <u>not</u> change radical

$8\sqrt{5}$

* If radicals are <u>not</u> the same, Simplify radicals. If radicals then match, add or subtract coeffecients and keep like radicals the same.

Step 1. Radicals must be the same

2. If radicals are **different**, simplify radicals if possible

3. Then, if radicals are the <u>same</u>, <u>add</u> coeffecients

80

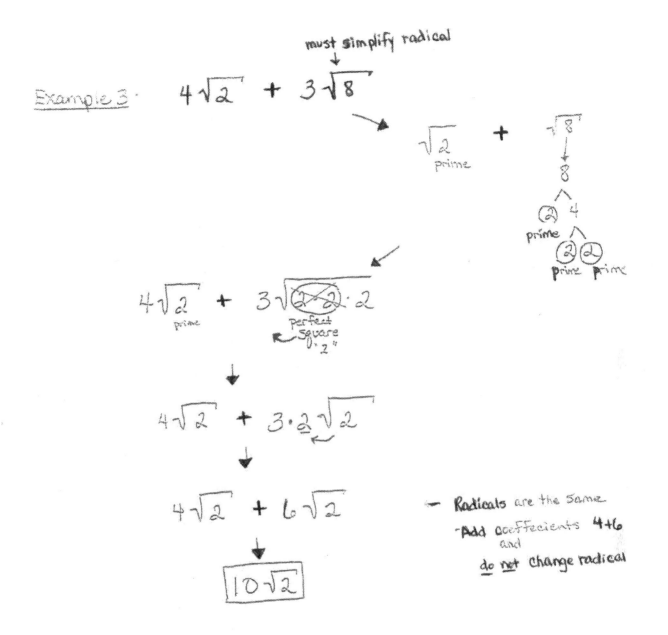

Example 3 $4\sqrt{2} + 3\sqrt{8}$

$\sqrt{2}$ prime   + $\sqrt{8}$

8

2 prime 4

2 prime 2 prime

$4\sqrt{2}$ prime + $3\sqrt{2 \cdot 2 \cdot 2}$

perfect square "2"

$4\sqrt{2}$ + $3 \cdot 2\sqrt{2}$

$4\sqrt{2}$ + $6\sqrt{2}$

– Radicals are the same
– Add coeffecients 4+6
  and
  do not change radical

$\boxed{10\sqrt{2}}$

MULTIPLYING RADICALS

※ Multiply all coeffecients (numbers in front of radical ($\sqrt{\ }$) sign). Then multiply all irrational numbers (numbers inside radical ($\sqrt{\ }$) signs). Then simplify if necessary.

Step 1. Multiply coeffecients
2. Multiply numbers inside radicals
3. Simplify radical using simplifying radicals rules if possible
4. Any perfect squares brought outside radical must be multiplied to the coeffecients.

Example 1 - $2\sqrt{3} \cdot 3\sqrt{5x}$

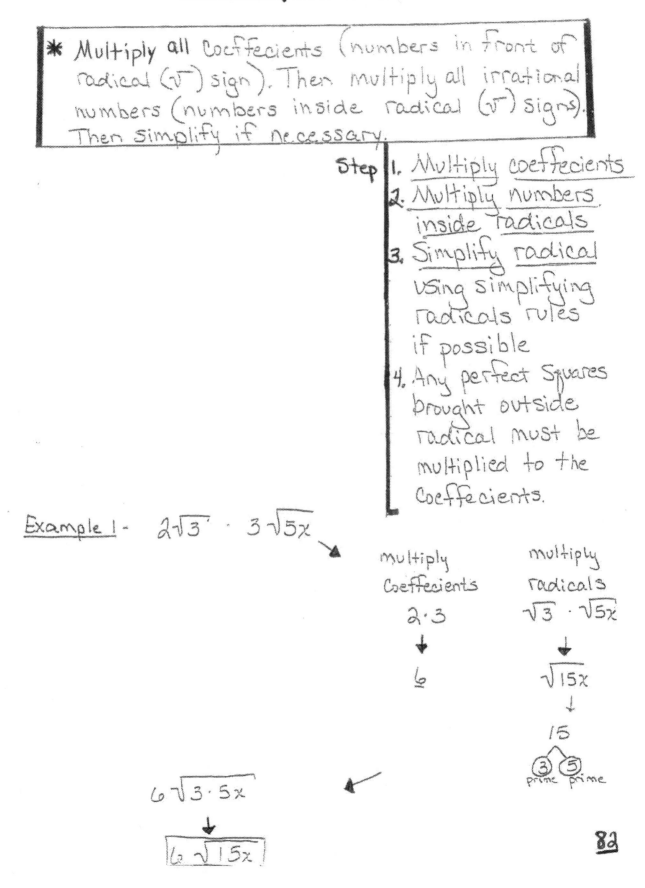

multiply coeffecients
$2 \cdot 3$
↓
6

multiply radicals
$\sqrt{3} \cdot \sqrt{5x}$
↓
$\sqrt{15x}$
↓
15
③ ⑤
prime prime

$6\sqrt{3 \cdot 5x}$
↓
$6\sqrt{15x}$

Example 2- $4\sqrt{6x}$ • $3\sqrt{15x^3}$

Mult
Coeffecients   Mult
                      Radicals
4·3 $6x^1 \cdot 15x^3$

↓   ↓
12   6   15
                     ② ③   ③ ⑤
                   prime prime prime prime

$12\sqrt{2 \cdot \cancel{3} \cdot \cancel{3} \cdot 5 \cdot \cancel{x} \cdot \cancel{x} \cdot \cancel{x} \cdot \cancel{x}}$
         perfect   perfect   perfect
         square   square   square
         "3"   "x"   "x"

$12 \cdot 3 \cdot x \cdot x \sqrt{2 \cdot 5}$

$$\boxed{36x^2\sqrt{10}}$$

Example 3 - $2\sqrt{7}$ • $4\sqrt{7}$

Mult   Mult
Coeffecients   Radicals

2·4   7·7
↓   ↓
8   49
                     ⑦   ⑦
                    prime   prime

$8\sqrt{\cancel{7} \cdot \cancel{7}}$
      perfect
      square
      "7"

$\boxed{56}$ ← $8 \cdot 7 \cancel{\sqrt{}}$ empty radical—
                                    throw away

83

RADICALS and INDEXES

A <u>radical number</u> is a number that needs to be prime factored and is under a radical sign.

An <u>**index**</u> tells how many of the <u>same</u> prime factors must be present to come out of a radical.

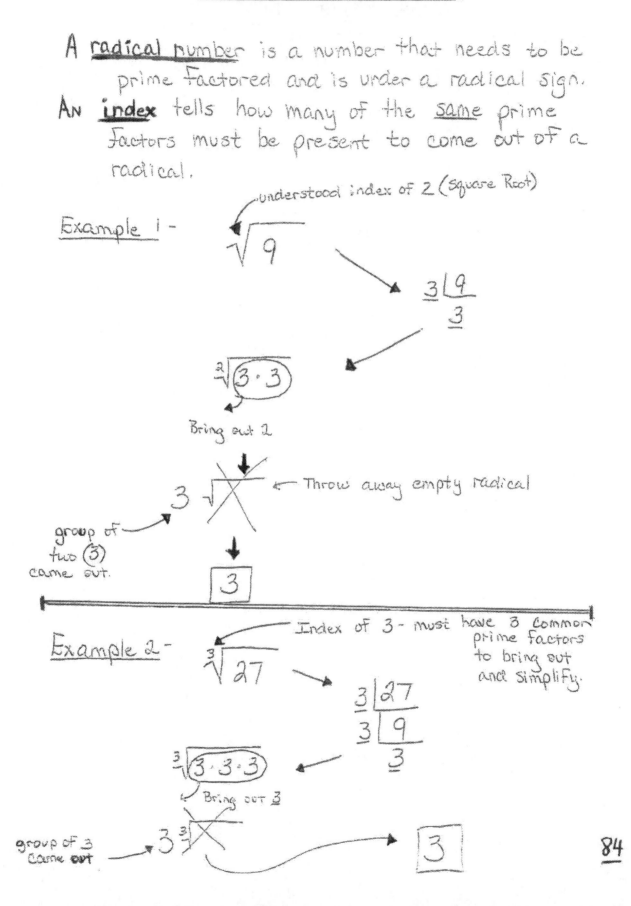

understood index of 2 (Square Root)

Example 1 -

$\sqrt{9}$

$3\lfloor\underline{9}$
$\qquad 3$

$\sqrt[2]{3 \cdot 3}$

Bring out 2

Throw away empty radical

$3\sqrt{}$

group of two ③ came out.

$\boxed{3}$

Example 2 -

Index of 3 - must have 3 common prime factors to bring out and simplify.

$\sqrt[3]{27}$

$3\lfloor\underline{27}$
$3\lfloor\underline{9}$
$\qquad 3$

$\sqrt[3]{3 \cdot 3 \cdot 3}$

Bring out 3

group of 3 came out

$3\sqrt[3]{}$

$\boxed{3}$

84

Example 3 -

$\sqrt[4]{32}$

Index 4 - **must have 4 common prime factors to bring out and simplify**

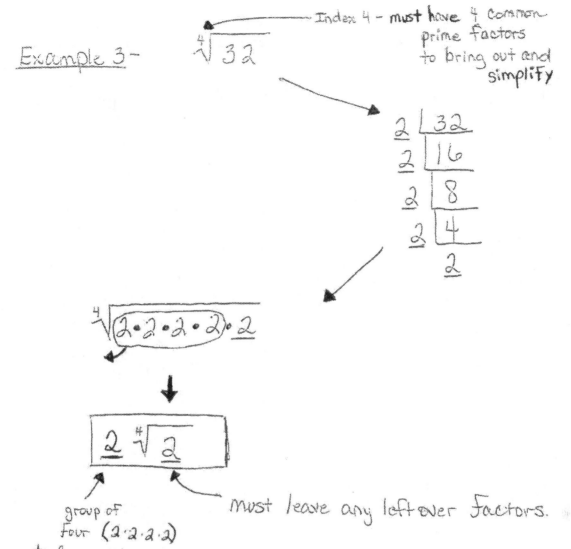

$$\begin{array}{c|c} 2 & 32 \\ 2 & 16 \\ 2 & 8 \\ 2 & 4 \\ & 2 \end{array}$$

$\sqrt[4]{2 \cdot 2 \cdot 2 \cdot 2 \cdot 2}$

$\boxed{2 \sqrt[4]{2}}$

group of four $(2 \cdot 2 \cdot 2 \cdot 2)$ to come out.

must leave any leftover factors.

85

RADICALS AND RATIONAL EXPONENTS

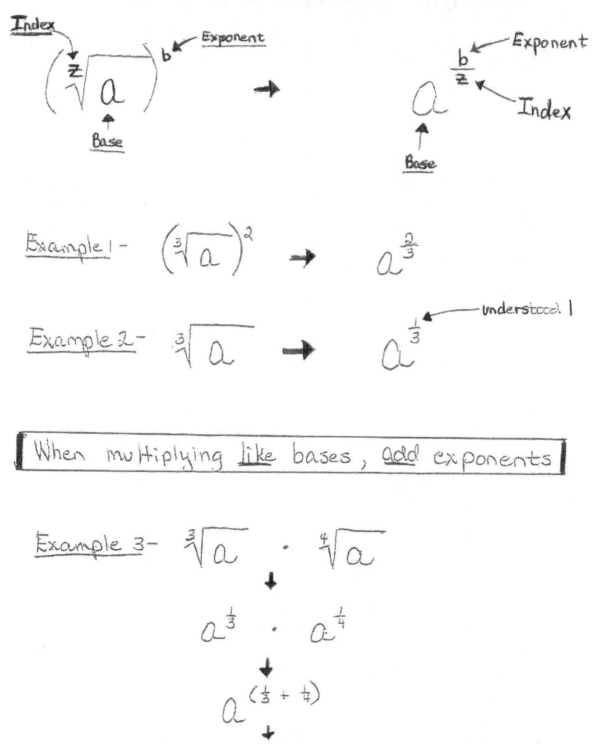

Index $\rightarrow$ $\left(\sqrt[N]{a}\right)^{b}$ $\leftarrow$ Exponent

Base

$\rightarrow$

Exponent $\rightarrow$ $a^{\frac{b}{N}}$ $\leftarrow$ Index

Base

Example 1- $\left(\sqrt[3]{a}\right)^{2}$ $\rightarrow$ $a^{\frac{2}{3}}$

Example 2- $\sqrt[3]{a}$ $\rightarrow$ $a^{\frac{1}{3}}$ $\leftarrow$ understood 1

When multiplying like bases, add exponents

Example 3- $\sqrt[3]{a} \cdot \sqrt[4]{a}$

$\downarrow$

$a^{\frac{1}{3}} \cdot a^{\frac{1}{4}}$

$\downarrow$

$a^{\left(\frac{1}{3} + \frac{1}{4}\right)}$

$\downarrow$

$a^{\frac{4}{12} + \frac{3}{12}}$

$\downarrow$

$\boxed{a^{\frac{7}{12}}}$

SIMPLIFYING RADICAL EXPRESSIONS

Prime factor radicals to simplify. Combine radicals by adding coeffecients if radicals are **exactly** the same.

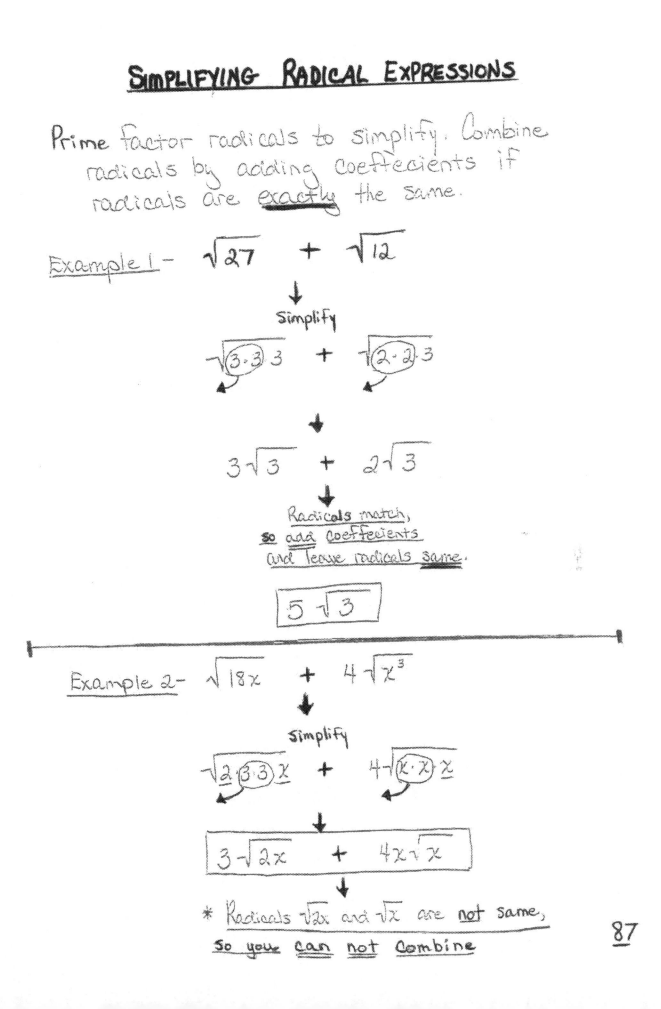

Example 1 - $\sqrt{27}$ + $\sqrt{12}$

↓

Simplify

$\sqrt{(3 \cdot 3) \, 3}$ + $\sqrt{(2 \cdot 2) \, 3}$

↓

$3\sqrt{3}$ + $2\sqrt{3}$

↓

Radicals match, so add coeffecients and leave radicals same.

$$\boxed{5\sqrt{3}}$$

Example 2 - $\sqrt{18x}$ + $4\sqrt{x^3}$

↓

Simplify

$\sqrt{2 \, (3 \cdot 3) \, x}$ + $4\sqrt{(x \cdot x) \, x}$

↓

$\boxed{3\sqrt{2x} \quad + \quad 4x\sqrt{x}}$

↓

* Radicals $\sqrt{2x}$ and $\sqrt{x}$ are **not** same, so you **can not** combine

87

SOLVING RADICAL EQUATIONS

Step

1. <u>Isolate</u> radical to one side of equation (remove anything added, subtracted or multiplied to radical)

2. Remove radical by squaring (if square root — cube if cube root, etc) <u>both</u> sides of equation.

<u>Example 1-</u>

$$\sqrt{2x+4} = 8$$

$$\left(\sqrt{2x+4}\right)^2 = (8)^2$$

$$2x+4 = 64$$
$$-4 \qquad -4$$

$$\frac{2x}{2} = \frac{60}{2}$$

$$\boxed{x = 30}$$

<u>Example 2-</u>

$$\sqrt{x+3} + 5 = 20$$
$$\phantom{\sqrt{x+3}}-5 \qquad -5$$

$$\sqrt{x+3} = 15$$

$$\left(\sqrt{x+3}\right)^2 = 15^2$$

$$x+3 = 225$$
$$-3 \qquad -3$$

$$\boxed{x = 222}$$

WHAT IS A QUADRATIC?

Quadratic- Lead variable is to the second power
(degree of 2)

*When graphed, it forms a shape called a
parabola ↯

* To solve a quadratic, set quadratic = 0
$$(x^2 = 0)$$

Quadratic Parent $y = x^2$

Quadratic Equation $y = ax^2 + bx + c$

QUADRATICS - BASICS OF GRAPHING

* <u>Quadratic</u> $y = x^2$ (<u>parent equation</u>)
    (parabola) $y = ax^2 + c$
                       (1)   (0)

<u>How quadratic moves</u>   <u>y-intercept</u>

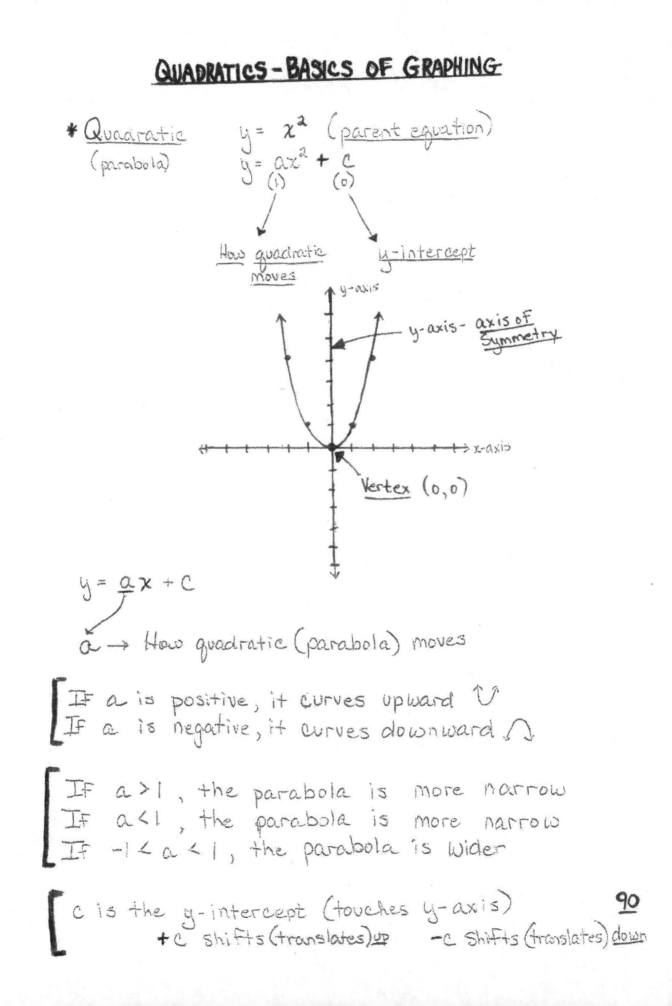

y-axis - <u>axis of symmetry</u>

Vertex (0,0)

$y = \underline{a}x + c$

$a \rightarrow$ How quadratic (parabola) moves

[If a is positive, it curves upward ∪
  If a is negative, it curves downward ∩

[If $a > 1$, the parabola is more narrow
  If $a < 1$, the parabola is more narrow
  If $-1 < a < 1$, the parabola is wider

[c is the y-intercept (touches y-axis)
     +c shifts (translates) <u>up</u>   -c shifts (translates) <u>down</u>

90

QUADRATIC FORMS

* <u>Standard Form</u> → $ax^2 + bx + \underline{c} = 0$

gives you <u>y-intercept</u>

<u>Example 1-</u> $x^2 - 4x - \underline{5} = 0$

y-intercept = $^-5$

---

* <u>Factored Form</u> → $(x \pm p)(x \pm q) = 0$

Set each = 0
and solve for x
to find <u>x-intercepts</u>

<u>Example 2 -</u>

$$(x - 5)(x + 1) = 0$$

$x - 5 = 0$ $x + 1 = 0$
  $+5 \ | \ +5$ $-1 \ | \ -1$

$\boxed{x = 5}$ $\boxed{x = -1}$

x-intercepts

---

* <u>Vertex Form</u> → $(x - h)^2 + k = 0$

Vertex
(h, k)

<u>Gives you vertex</u>

<u>Example 3 -</u> $(x - 2)^2 - 9 = 0$

$(2, -9)$ ← <u>Vertex</u>
 h, k

91

Graph from

Standard form → $x^2 - 4x - 5 = 0$

y-intercept
-5

Factored form → $(x - 5)(x + 1) = 0$

+5   -1

x-intercepts

Vertex form → $(x - 2)^2 - 9 = 0$

$(2, -9)$

vertex
(point parabola turns)
Highest or lowest point

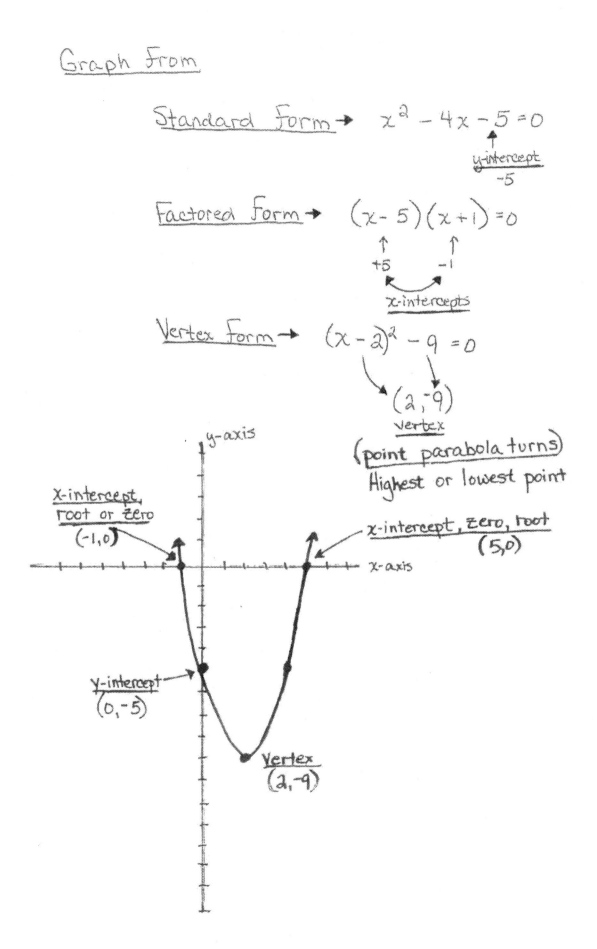

x-intercept,
root or zero
$(-1, 0)$

x-intercept, zero, root
$(5, 0)$

y-axis

x-axis

y-intercept
$(0, -5)$

Vertex
$(2, -9)$

QUADRATIC FORMULA

* <u>Quadratic Formula</u> – used to find roots or solutions of a quadratic graph or equation (where a quadratic graph touches or passes through the x-axis or solves for x when y = 0)

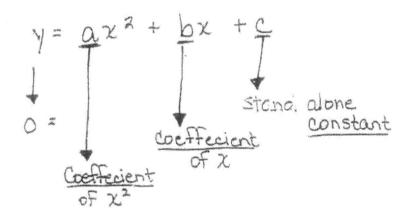

$$y = ax^2 + bx + c$$

$$0 =$$

<u>Coeffecient</u>
of x^2

<u>Coeffecient</u>
of x

Stand alone
<u>constant</u>

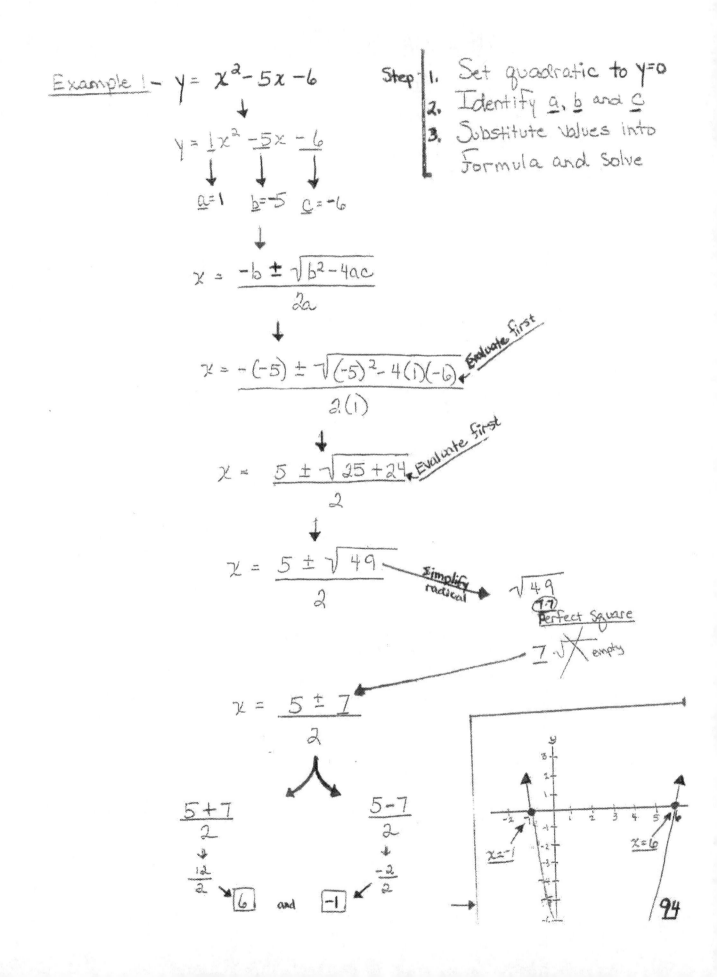

Example 1 - $y = x^2 - 5x - 6$

$y = 1x^2 - 5x - 6$

$a = 1$ $b = -5$ $c = -6$

$$x = \frac{-b \pm \sqrt{b^2 - 4ac}}{2a}$$

$$x = \frac{-(-5) \pm \sqrt{(-5)^2 - 4(1)(-6)}}{2(1)}$$ Evaluate first

$$x = \frac{5 \pm \sqrt{25 + 24}}{2}$$ Evaluate first

$$x = \frac{5 \pm \sqrt{49}}{2}$$ Simplify radical

$\sqrt{49}$
7·7 Perfect Square
7 $\sqrt{}$ empty

$$x = \frac{5 \pm 7}{2}$$

$\dfrac{5 + 7}{2}$ $\dfrac{5 - 7}{2}$

$\dfrac{12}{2}$ → $\boxed{6}$ and $\boxed{-1}$ ← $\dfrac{-2}{2}$

Step
1. Set quadratic to y=0
2. Identify a, b and c
3. Substitute values into formula and Solve

x = -1   x = 6

94

Example 2 - $y = x^2 - 4x + 4$

$$\downarrow$$

$y = 1x^2 - 4x + 4$

$\downarrow$ $\downarrow$ $\downarrow$ $c=4$

$a=1$ $b=-4$

$$\downarrow$$

$$x = \frac{-b \pm \sqrt{b^2 - 4ac}}{2a}$$

$$\downarrow$$

$$x = \frac{-(-4) \pm \sqrt{(-4)^2 - 4(1)(4)}}{2(1)}$$ ← Evaluate first

$$\downarrow$$

$$x = \frac{4 \pm \sqrt{16 - 16}}{2}$$ ← Evaluate first

$$\downarrow$$

$$x = \frac{4 \pm \sqrt{0}}{2} \longrightarrow \sqrt{0} = 0$$

$$x = \frac{4 \pm 0}{2}$$

$\frac{4+0}{2}$ $\frac{4-0}{2}$

$\downarrow$ $\downarrow$

$\frac{4}{2} \rightarrow \boxed{2}$ and $\boxed{2} \leftarrow \frac{4}{2}$

x=2

Example 3. $y = 2x^2 - 5x + 4$

$$\downarrow$$

$$0 = 2x^2 - 5x + 4$$

$$a = 2 \qquad b = -5 \qquad c = 4$$

$$\downarrow$$

$$x = \frac{-b \pm \sqrt{b^2 - 4ac}}{2a}$$

$$\downarrow$$

$$x = \frac{-(-5) \pm \sqrt{(-5)^2 - 4(2)(4)}}{2(2)}$$

$$\downarrow$$

$$x = \frac{5 \pm \sqrt{25 - 32}}{4}$$

$$\downarrow$$

$$x = \frac{5 \pm \sqrt{-7}}{4}$$

$$\downarrow$$

$$x = \frac{5 \pm i\sqrt{7}}{4}$$

$\sqrt{-}$ negative radicals are imaginary numbers $\sqrt{-} \rightarrow i\sqrt{}$

$$\frac{5 + i\sqrt{7}}{4} \qquad\qquad \frac{5 - i\sqrt{7}}{4}$$

Imaginary Answers —
Do not touch x-axis

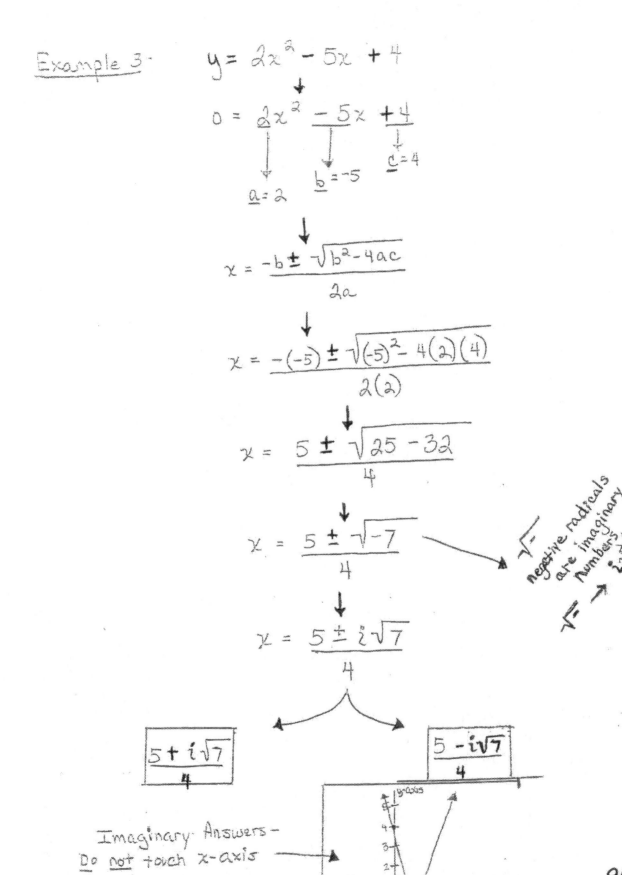

96

Example 4 -

$$x^2 - 4x = 8$$

$$\downarrow$$

$$x^2 - 4x = 8$$
$$ -8 -8$$
$$\overline{x^2 - 4x - 8 = 0}$$

set quadratics $= 0$

$$\downarrow$$

$$1x^2 - 4x - 8 = 0$$

$$\downarrow \downarrow \downarrow$$
$$ c = -8$$

$$a=1 b = -4$$

$$\downarrow$$

$$x = \frac{-b \pm \sqrt{b^2 - 4ac}}{2a}$$

$$\downarrow$$

$$x = \frac{-(-4) \pm \sqrt{(-4)^2 - 4(1)(-8)}}{2(1)}$$

$$\downarrow$$

$$x = \frac{4 \pm \sqrt{16 + 32}}{2}$$

$$\downarrow$$

$$x = \frac{4 \pm \sqrt{48}}{2} \longrightarrow$$

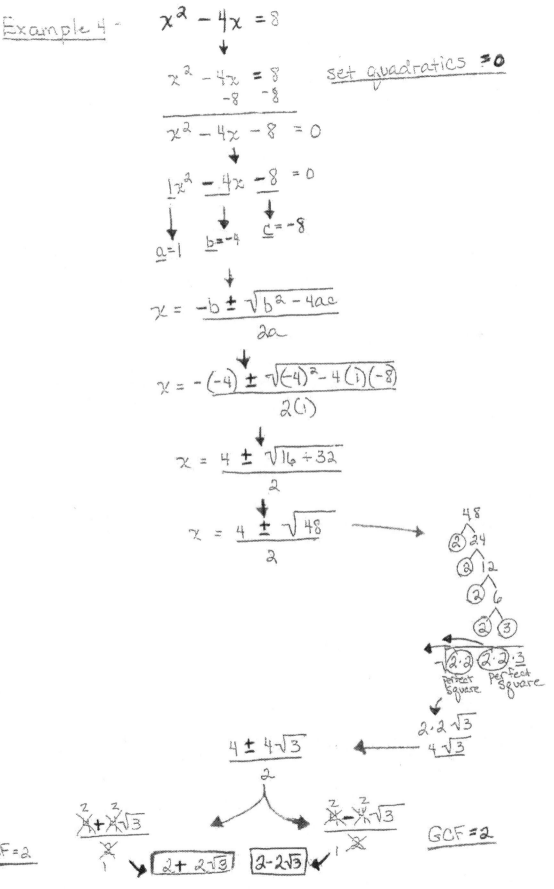

$$2 \cdot 2\sqrt{3}$$
$$4\sqrt{3}$$

$$\frac{4 \pm 4\sqrt{3}}{2}$$

$$\frac{\cancel{4}^2 + \cancel{4}^2\sqrt{3}}{\cancel{2}_1} \frac{\cancel{4}^2 - \cancel{4}^2\sqrt{3}}{\cancel{2}_1}$$

GCF = 2

$$\boxed{2 + 2\sqrt{3}} \boxed{2 - 2\sqrt{3}}$$

GCF = 2

97

DISCRIMINANT

* Discriminant - value gives information about the roots of a quadratic equation or graph

* Using the quadratic equation -

Discriminant

$$\frac{-b \pm \sqrt{b^2 - 4ac}}{2a}$$

Discriminant
$$b^2 - 4ac$$

* The discriminant quickly determines how many solutions (roots) a quadratic

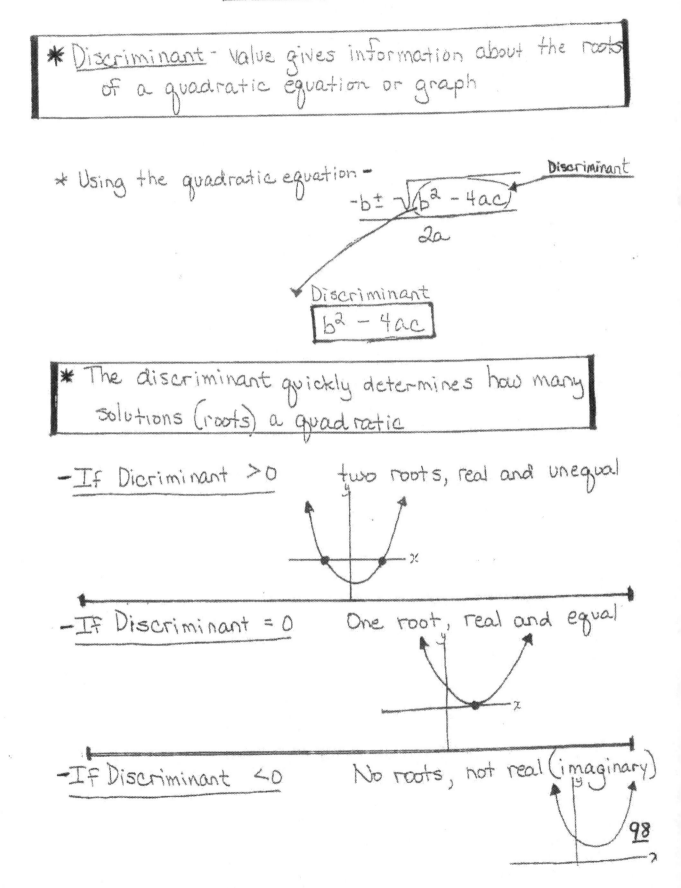

- If Discriminant > 0   two roots, real and unequal

- If Discriminant = 0   One root, real and equal

- If Discriminant < 0   No roots, not real (imaginary)

Example 1- $x^2 + 5x + 4 = 0$

$1x^2 + 5x + 4 = 0$

$a = 1$
$b = 5$
$c = 4$

Discriminant

$b^2 - 4ac$

$(5)^2 - 4(1)(4)$

$25 - 16$

$\boxed{9}$ Two, real, unequal Solutions

Example 2- $5x^2 + 3x + 1 = 0$

$a = 5$
$b = 3$
$c = 1$

Discriminant

$b^2 - 4ac$

$(3)^2 - 4(5)(1)$

$9 - 20$

$\boxed{-11}$ → No roots, imaginary Solutions

Example 3- $x^2 - 6x + 9 = 0$

$1x^2 - 6x + 9 = 0$

$a = 1$
$b = -6$
$c = 9$

Discriminant

$b^2 - 4ac$

$(-6)^2 - 4(1)(9)$

$36 - 36$

one real solution $\boxed{0}$

99

FACTORING

* <u>2 Terms</u>   <u>Either</u> → * Factor out common factors - if possible

or

* Difference of 2 perfect squares (or sum/difference of two perfect cubes)

---

* <u>Factoring out common factors:</u>

<u>Example 1</u> - $4x^2 + 10x$

<u>Common factor of 2</u>
and since they both have "x"
in common, smallest "x" is $\underline{x^1}$

$\boxed{2x^1(2x + 5)}$

<u>Example 2</u> - $15x^3 - 5x^2$

<u>Common factor of 5</u>
and since the both have
"x" in common, smallest
"x" is $\underline{x^2}$

$\boxed{5x^2(3x - 1)}$

<u>Step</u> 1. Factor out greatest common factor (GCF) of coeffecients (if possible)

2. Factor out common variables - factor out each variable with smallest exponent

3. Keep in parenthesis all remaining coeffecients and variables <u>not</u> factored out - if any

<u>Remember</u> - if you factor out entire term, you <u>**must**</u> replace the term in parenthesis with $\underline{1}$

100

Example 3

$$6xy^3z^8 + 14x^3y^2$$

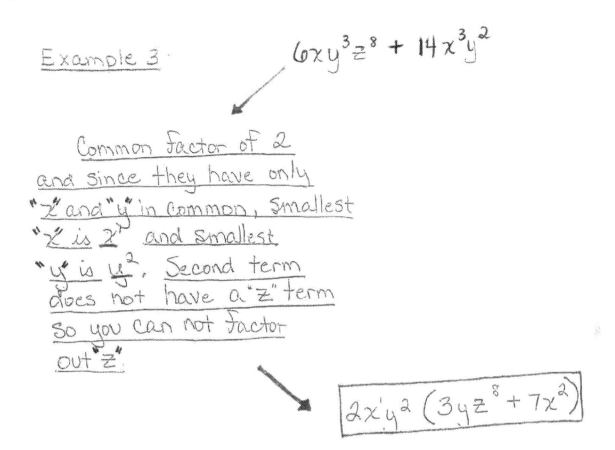

Common factor of 2
and since they have only
"z" and "y" in common, smallest
"x" is x^1 and smallest
"y" is y^2. Second term
does not have a "z" term
so you can not factor
out "z".

$$2x^1y^2(3yz^8 + 7x^2)$$

*Difference of 2 Perfect Squares:

Step 1. Must be difference, (not sum)

2. Each of the two terms must be a perfect square (coefficients, variables, and constants)

3. Each variables' exponent __must__ be even - Square root is __half__ the exponent

4. Put square roots into two seperate parenthesis

$$(\underline{\quad} + \underline{\quad})(\underline{\quad} - \underline{\quad})$$

Square roots of first term — Square roots of second term

Example 1 -

$$x^2 - 49$$

$$x^2 \quad - \quad 49$$

Even exponent - perfect Square $x \cdot x$

Difference

Perfect Square $7 \cdot 7$

$$\boxed{(x+7)(x-7)}$$

Example 2 -

$$x^6 - 16$$

$$x^6 \quad - \quad 16$$

Even exponent perfect Square $x^3 \cdot x^3$

difference

Perfect Square $4 \cdot 4$

$$\boxed{(x^3 + 4)(x^3 - 4)}$$

Example 3 -

$$9x^2 - 100y^4$$

Even exponent both perfect squares $x \cdot x$ and $3 \cdot 3$

Even exponent both perfect squares $y^2 \cdot y^2$ and $10 \cdot 10$

$$\boxed{(3x + 10y^2)(3x - 10y^2)}$$

102

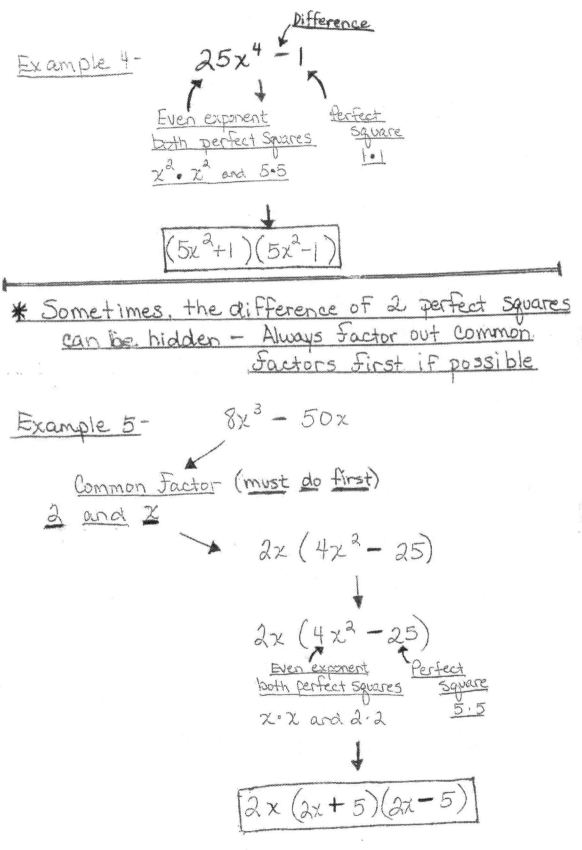

Example 4-

Difference

$25x^4 - 1$

Even exponent
both perfect squares
$x^2 \cdot x^2$ and $5 \cdot 5$

Perfect square
$1 \cdot 1$

$$(5x^2 + 1)(5x^2 - 1)$$

✱ Sometimes, the difference of 2 perfect squares can be hidden — Always factor out common factors first if possible

Example 5- $8x^3 - 50x$

Common factor (must do first)
2 and x

$2x(4x^2 - 25)$

$2x(4x^2 - 25)$

Even exponent
both perfect squares
$x \cdot x$ and $2 \cdot 2$

Perfect square
$5 \cdot 5$

$$2x(2x + 5)(2x - 5)$$

* When factoring out common factors, always make sure the term with the greatest exponent (lead term - highest degree) is <u>positive</u>. If it is negative, always factor out a (–) negative as well as common factors

Example 6- $-2x^3 + 32x$

$\downarrow$

$-2x^3 + 32x$

x^3 <u>must be positive</u> so factor out a <u>negative</u> common factor

$\underline{2}$ and $\underline{x}$

$-2x(x^2 - 16)$

<u>Even exponent</u>   <u>Perfect square</u>
$x \cdot x$ $4 \cdot 4$

$\downarrow$

$\boxed{-2x(x+4)(x-4)}$

*3 Terms - Quadratic (Trinomial)

* Last term is positive (+)

Step 1. Factor out all common factors (Same as factoring 2-Terms)

2. Identify sign in front of last term - If positive (+), each parenthesis will have the Same signs → Either

$$(+)(+)$$

or

$$(-)(-)$$

The sign in the middle of the problem will tell you which ones to use

3. Find all factors of last term

4. Find the set of factors that **add** up to middle term

Example 1—

$$x^2 + 5x + 6$$

↓

$$x^2 + 5x + 6$$

Signs same

Both signs
⊕

Step 2

↓

$$(+)(+)$$

↓

$$x^2 + 5x + 6$$

↓

Factors
1·6
(2·3)
↓
add up to
5

Step 3

Step 4

↓

$$\boxed{(x+2)(x+3)}$$

So 2·3=6
2+3=5

$$x^2 + 5x + 6$$

105

Example 2- $x^2 - 7x + 6$

$\downarrow$

$x^2 - 7x + 6$

$\underline{\text{Signs same}}$

Both Signs   Step2
$\underline{\quad\quad\quad}$
$\ominus$

$\downarrow$

$(\quad - \quad)(\quad - \quad)$

$\downarrow$

$x^2 - 7x + 6$

$\downarrow$

$\underline{\text{Factors}}$ Step 3
$\boxed{\text{1 • 6}}$
2 • 3

$\downarrow$

$\underline{\text{add up to}}$
$\underline{7}$ Step 4

So 1 • 6 = 6
and
1 + 6 = 7

$\boxed{(x - 1)(x - 6)}$ $x^2 - 7x + 6$

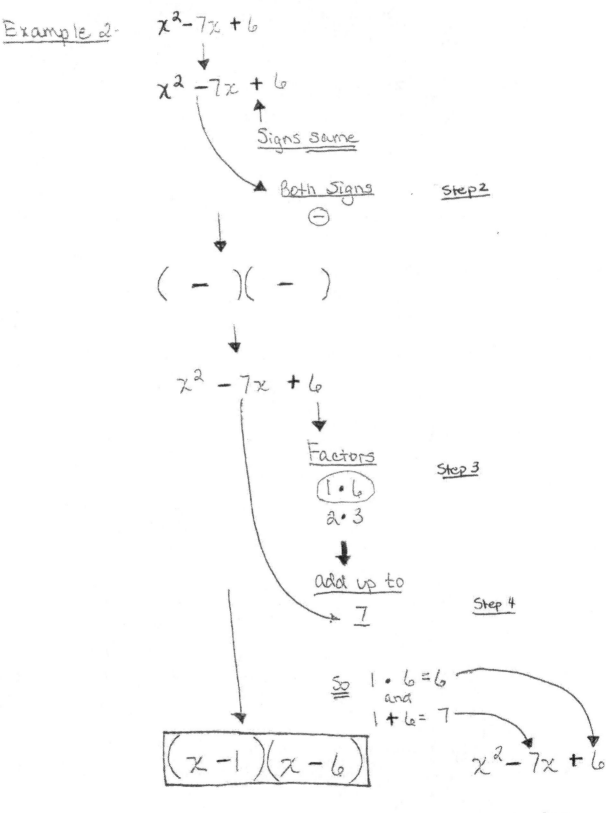

* Last term is negative (−)

Example 3- $x^2 + 5x - 6$

$$\downarrow$$

$$x^2 + 5x - 6$$

Signs are different

$$(\ + \ )(\ - \ )$$

Step 2

$$x^2 + 5x - 6$$

Step 3

Factors
(1 · 6)
2 · 3

$$\downarrow$$

Subtracted
5

Largest
Factor (+)

$$(x + 6)(x - 1)$$

Largest
Factor
6

$$\boxed{(x + 6)(x - 1)}$$

Step 1. Factor out all common factors (Same as Factoring 2-Terms)

2. Identify sign in front of last term — IF negative (−), each parenthesis will have different signs →
$$(\ + \ )(\ - \ )$$

3. Find all factors of last term

4. Find which set of factors subtracted gives the middle term
$$\downarrow$$
Largest of two factors goes with sign in the middle.

SD
$$1 \cdot 6 = 6$$
$$6 - 1 = 5$$

$$x^2 + 5x - 6$$

* 4 Terms - Factor by Grouping

Example 1 - $2x^3 + 8x^2 + 3x + 12$

$$2x^3 + 8x^2 \mid + 3x + 12$$

$$2x^2(x+4) \quad +3(x+4)$$

$$\underline{2x^2}(x+4) \qquad \underline{+3}(x+4)$$

Same

drop $(x+4)$ down

Step 5

$$(2x^2+3)(x+4)$$

$$\boxed{(2x^2+3)(x+4)}$$

Step 1. Factor out all common factors - if possible

2. Divide problem in half

Step 2   3. In first 2 terms, factor out all common factors

4. In last 2 terms factor out all common factors

Step 3 and 4

* On both sides, what is left in <u>both</u> parenthesis should be the <u>same</u>

5. Join together in a second parenthesis both terms factored out and copy next to what was left in both parenthesis

Example 3:

$$5x^3 + 10x - 2x - 4$$

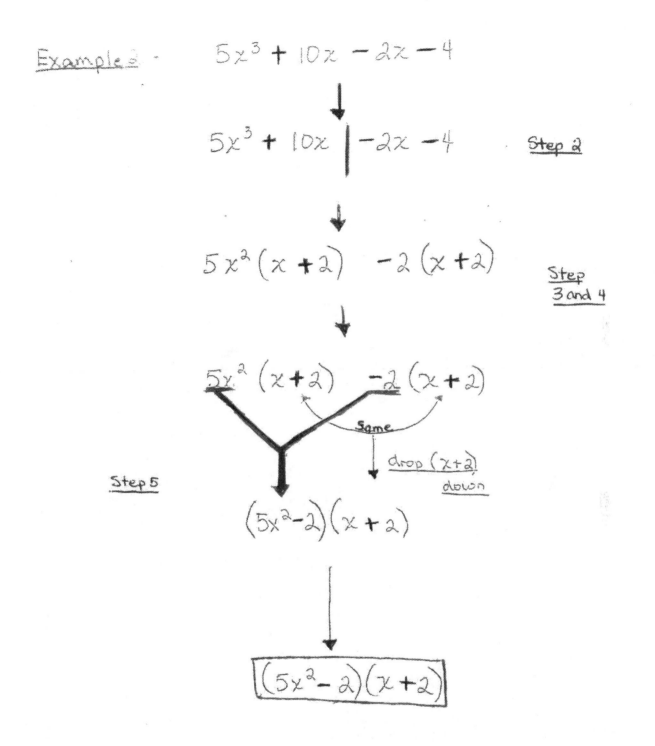

$$5x^3 + 10x \mid -2x - 4$$ Step 2

$$5x^2(x+2) \quad -2(x+2)$$ Step 3 and 4

$$5x^2(x+2) \quad -2(x+2)$$

Same

drop (x+2) down

Step 5

$$(5x^2-2)(x+2)$$

$$\boxed{(5x^2-2)(x+2)}$$

109

SOLVING QUADRATICS

> If equation has both an x^2 and an x, then set equation $= 0$.

Example 1 — $x^2 + 3x - 10$

set $= 0$ to solve

$$x^2 + 3x - 10 = 0$$

Factor

$$(x + 5)(x - 2) = 0$$

Set $= 0$ Set $= 0$

$x + 5 = 0$ $x - 2 = 0$
$\quad -5 \quad -5$ $\quad +2 \quad +2$

$\boxed{x = -5}$ $\boxed{x = 2}$

Example 2 — $x^2 = x + 6$

Set $= 0$

$$\begin{array}{rcl} x^2 & = & x + 6 \\ -x - 6 & & -x \quad -6 \end{array}$$

$$x^2 - x - 6 = 0$$

Factor

$$(x - 3)(x + 1) = 0$$

Set $= 0$ Set $= 0$

$x - 3 = 0$ $x + 1 = 0$
$\quad +3 \quad +3$ $\quad -1 \quad -1$

$\boxed{x = 3}$ $\boxed{x = -1}$

110

If there is an x^2 and **NO** x in the equation, then isolate x^2 and take $\sqrt{}$ to both side ($\pm$ to non-x side)

Example 3 - $x^2 - 9 = 0$

isolate x^2

$$x^2 - 9 = 0$$
$$ + 9 + 9$$

$$x^2 = 9$$

take $\sqrt{}$ both sides

$$\sqrt{x^2} = \pm\sqrt{9}$$

$3\lvert\frac{9}{3}$

$$\boxed{x = \pm 3}$$

Example 4 - $x^2 - 7 = 0$

isolate x

$$x^2 - 7 = 0$$
$$ + 7 + 7$$

$$x^2 = 7$$

take $\sqrt{}$ both sides

$$\sqrt{x^2} = \pm\sqrt{7}$$

can not simplify $\sqrt{7}$

$$\boxed{x = \pm\sqrt{7}}$$

MULTIPLYING POLYNOMIALS

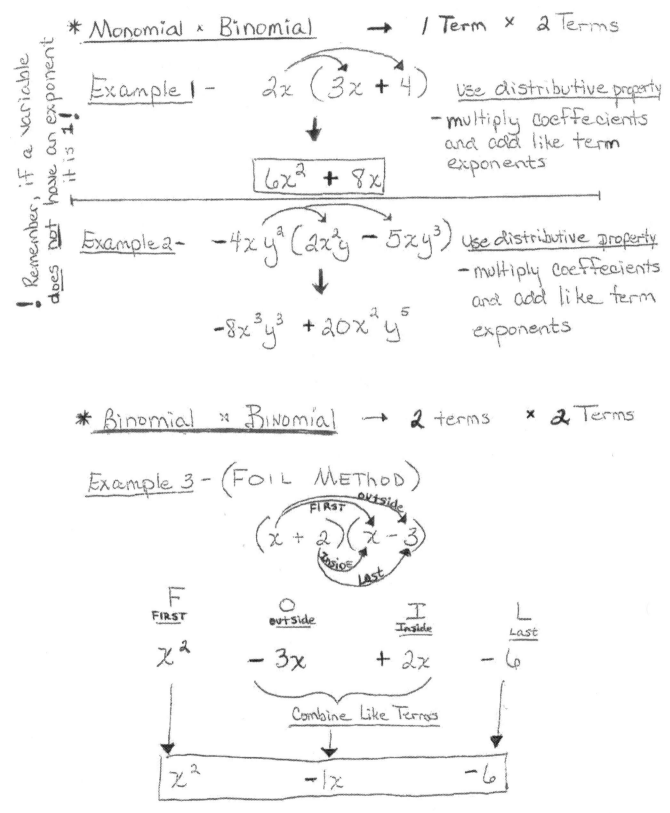

* <u>Monomial</u> × <u>Binomial</u> → 1 Term × 2 Terms

Example 1 - $2x(3x + 4)$ <u>Use distributive property</u>

– multiply coeffecients and add like term exponents

$$6x^2 + 8x$$

Example 2 - $-4xy^2(2x^2y - 5xy^3)$ <u>Use distributive property</u>

– multiply coeffecients and add like term exponents

$$-8x^3y^3 + 20x^2y^5$$

! Remember, if a variable does <u>not</u> have an exponent it is 1 !

* <u>Binomial × Binomial</u> → 2 terms × 2 Terms

Example 3 - (FOIL METHOD)

$(x + 2)(x - 3)$

FIRST outside INSIDE LAST

F FIRST	O outside	I Inside	L Last
x^2	$-3x$	$+2x$	-6

Combine Like Terms

x^2	$-1x$	-6

Example 4 - (Box Method)

$$(x+4)(x-6)$$

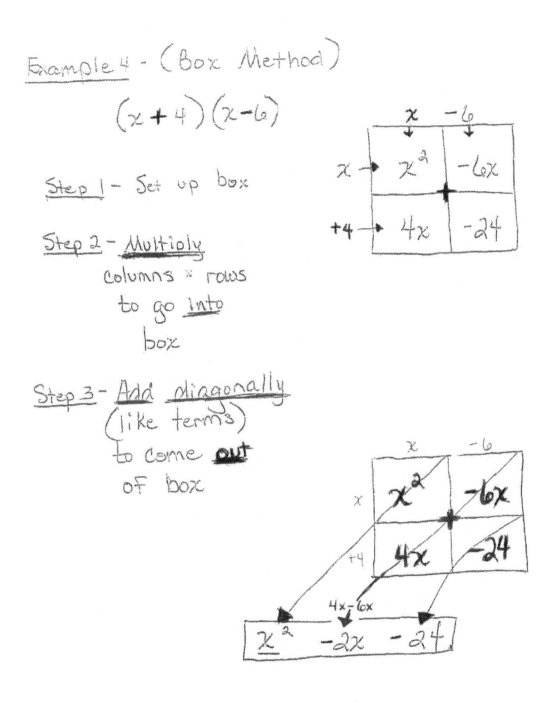

Step 1 - Set up box

Step 2 - <u>Multiply</u>
columns × rows
to go <u>into</u>
box

Step 3 - <u>Add</u> <u>diagonally</u>
(like terms)
to come **out**
of box

DIVIDING POLYNOMIALS

Divide each term on top of fraction bar by the bottom term — Reduce coeffecients of each new term and use division of exponent rules with each term.

Example 1-
$$\frac{12x^5 - 6x^3 + 4x^2}{4x^2}$$

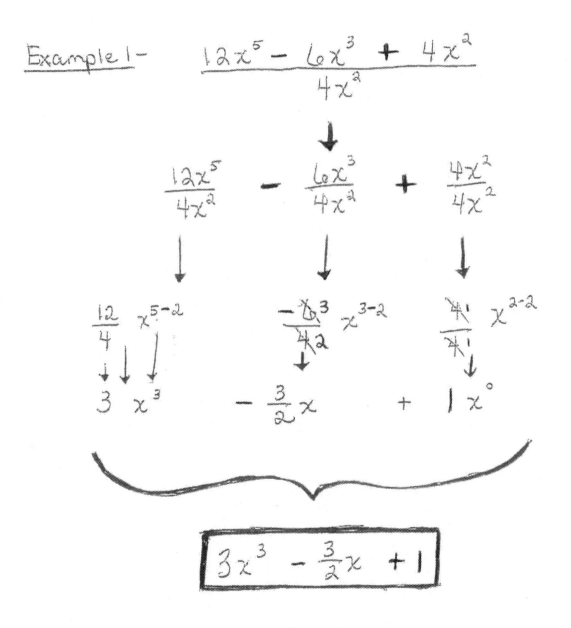

$$\frac{12x^5}{4x^2} \quad - \quad \frac{6x^3}{4x^2} \quad + \quad \frac{4x^2}{4x^2}$$

$$\frac{12}{4} x^{5-2} \qquad -\frac{6^3}{4_2} x^{3-2} \qquad \frac{4^1}{4^1} x^{2-2}$$

$$3 \ x^3 \qquad\qquad -\frac{3}{2}x \qquad\qquad + \ 1 x^0$$

$$\boxed{3x^3 - \frac{3}{2}x + 1}$$

IMAGINARY NUMBERS

* An <u>Imaginary Number</u> is a negative square root $(\sqrt{-})$ Remove the negative sign from inside the square root and call it imaginary (i)

Step 1. Remove − sign from inside radical $(\sqrt{-})$ and bring it outside as i

2. Simplify radical if possible as usual with i in front of radical or behind any rational number

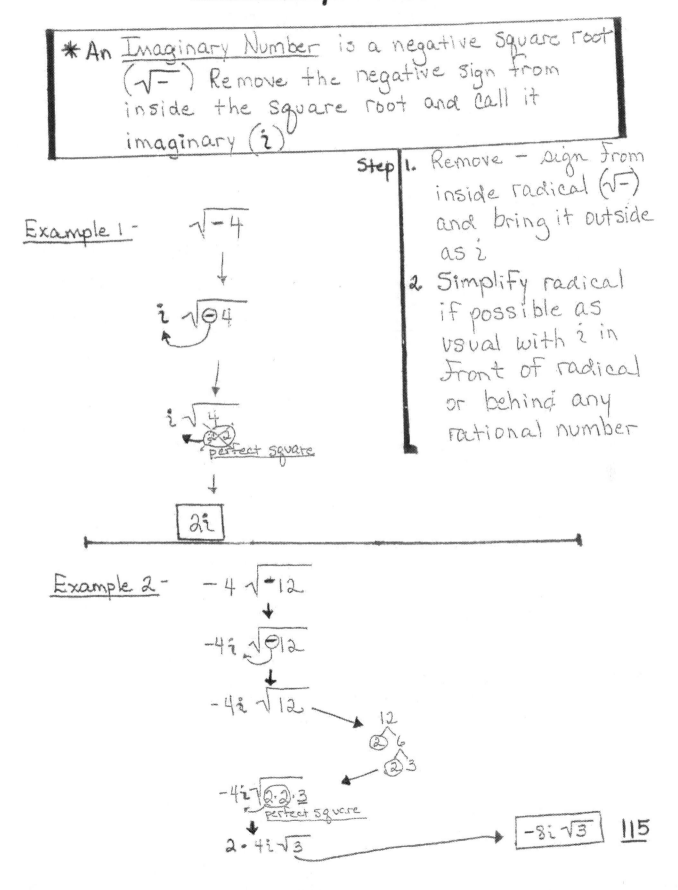

Example 1 - $\sqrt{-4}$

$\downarrow$

$i \sqrt{-4}$

$\downarrow$

$i \sqrt{4}$
   perfect square

$\downarrow$

$\boxed{2i}$

Example 2 - $-4\sqrt{-12}$

$\downarrow$

$-4i\sqrt{-12}$

$\downarrow$

$-4i\sqrt{12}$ → 12
                  2 6
                     2 3

$-4i\sqrt{2 \cdot 2 \cdot 3}$
      perfect square

$\downarrow$

$2 \cdot 4i\sqrt{3}$ → $\boxed{-8i\sqrt{3}}$ 115

ADDITION AND SUBTRACTION OF IMAGINARY NUMBERS

✱ <u>Imaginary Numbers</u> can only be added and subtracted to other <u>imaginary numbers</u> — if all numbers being added or subtracted are <u>all</u> imaginary, then <u>add</u> their coeffecients and label answer as imaginary (i).

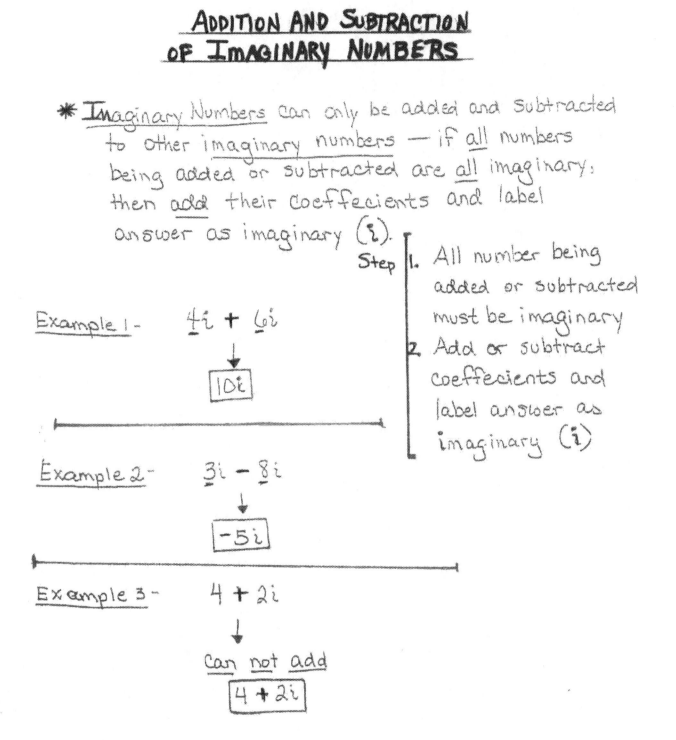

Step 1. All number being added or subtracted must be imaginary

2. Add or subtract coeffecients and label answer as imaginary (i)

<u>Example 1</u> - $\underline{4}i + \underline{6}i$

↓

$\boxed{10i}$

<u>Example 2</u> - $\underline{3}i - \underline{8}i$

↓

$\boxed{-5i}$

<u>Example 3</u> - $4 + 2i$

↓

<u>Can not add</u>

$\boxed{4 + 2i}$

MULTIPLYING IMAGINARY NUMBERS

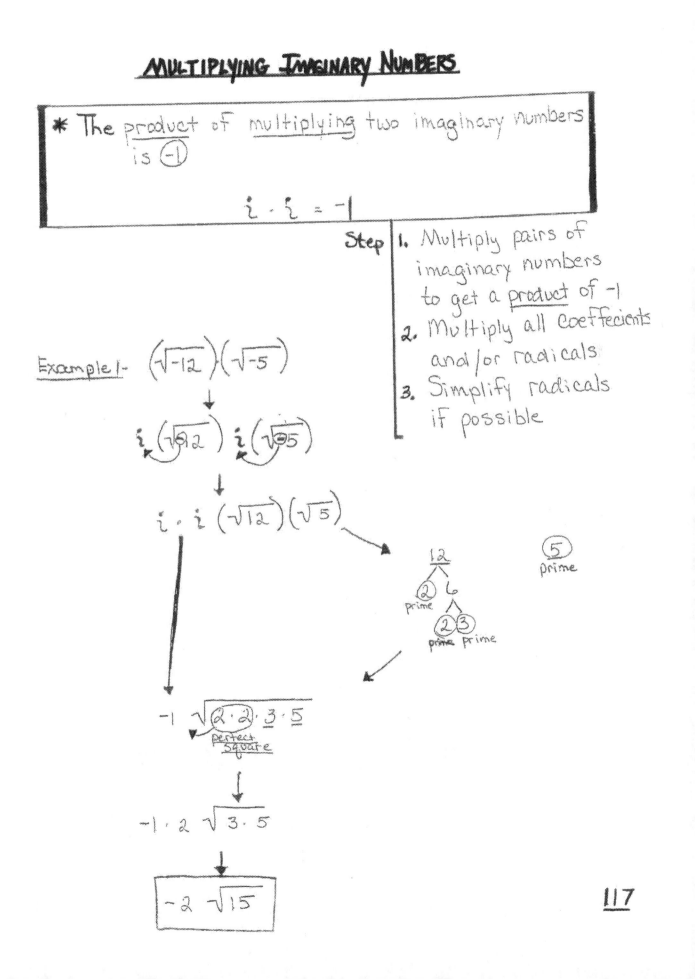

* The product of multiplying two imaginary numbers is (-1)

$$i \cdot i = -1$$

Step 1. Multiply pairs of imaginary numbers to get a product of -1
2. Multiply all coeffecients and/or radicals
3. Simplify radicals if possible

Example 1- $(\sqrt{-12})(\sqrt{-5})$

$i(\sqrt{-12})\ i(\sqrt{-5})$

$i \cdot i\ (-\sqrt{12})(\sqrt{5})$

12
2 prime   6
         2 prime 3 prime

5 prime

$-1 \sqrt{2 \cdot 2 \cdot 3 \cdot 5}$
perfect square

$-1 \cdot 2 \sqrt{3 \cdot 5}$

$-2 \sqrt{15}$

117

Example 2- $\left(i\sqrt{12} \right)\left(i\sqrt{3} \right)$

$\downarrow$

$i \cdot i \sqrt{12} \cdot \sqrt{3}$

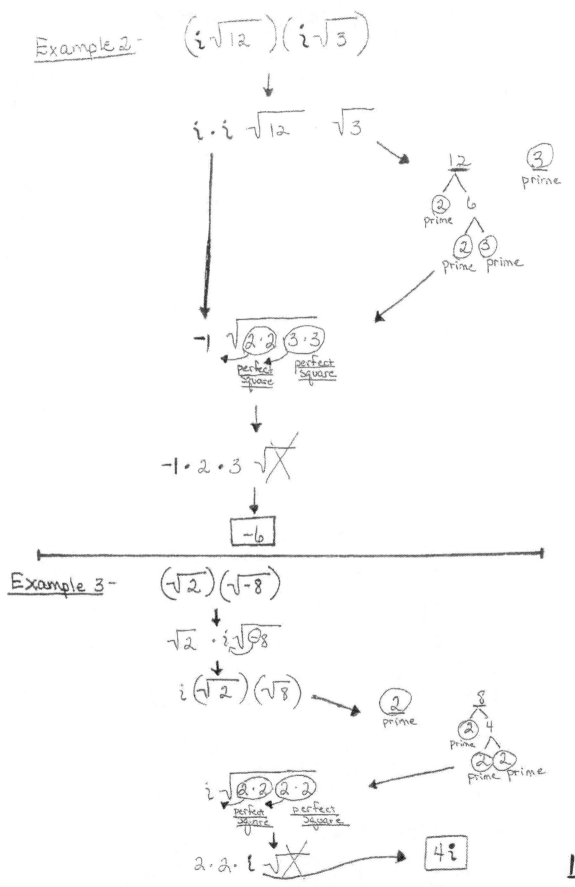

12
⑵ 6
prime
⑵ ③
prime prime

③
prime

$\neg 1 \sqrt{(2\cdot 2)(3\cdot 3)}$
perfect
square   perfect
square

$\downarrow$

$-1 \cdot 2 \cdot 3 \sqrt{}$

$\downarrow$

$\boxed{-6}$

Example 3- $\left(\sqrt{2} \right)\left(\sqrt{-8} \right)$

$\downarrow$

$\sqrt{2} \cdot i\sqrt{8}$

$\downarrow$

$i\left(\sqrt{2} \right)\left(\sqrt{8} \right)$

②
prime

8
⑵ 4
prime
⑵⑵
prime prime

$i\sqrt{(2\cdot 2)(2\cdot 2)}$
perfect
square   perfect
square

$\downarrow$

$2 \cdot 2 \cdot i \sqrt{}$ → $\boxed{4i}$

CONJUGATES

* A <u>conjugate</u> is formed by changing the sign between two terms in a binominal. Conjugates are used to remove radical or imaginary numbers in a denominator.

* The product of 2 conjugates is always a rational number - <u>Rationalizing the Denominator</u>

<u>Example 1-</u>   What is the conjugate of $(x+4)$?

$$(x+4)\underline{\underline{(x-4)}}$$

<u>why?</u>

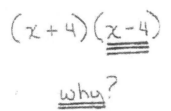

$(x+4)(x-4)$

F First	O outside	I Inside	L Last
x^2	$-4x$	$+4x$	-16

x^2 $\cancel{-4x}$ $\cancel{+4x}$ -16
<u>cancel out</u>

$$\boxed{x^2 - 16}$$

119

RATIONAL EXPRESSIONS

* <u>Rational Expression</u>- both numerator and denominator are polynomials. To simplify a rational expression, eliminate all factors that are common in the numerator and denominator.

Example 1-

$$\frac{x+5}{x}$$

$\downarrow$

$$\frac{(x+5)}{x}$$

← Complex numbers in a fraction must be put in a parenthesis ($x+5$ is one number)

$\downarrow$

$$\frac{(x+5)}{x}$$ can not simplify

$\downarrow$

$$\boxed{\frac{x+5}{x} \quad x \neq 0}$$

Step 1. Factor numerator and denominator if possible

2. Cancel out all common factors (GCF)

3. Since denominators <u>can not</u> be zero (0 in the denominator is an undefined number), list any excluded domain values that make answer undefined.

If $x=0$, then it would not be a real number (Denominators <u>can not</u> be $=0$)

120

Example 2 - $\dfrac{8ab}{12b^2}$

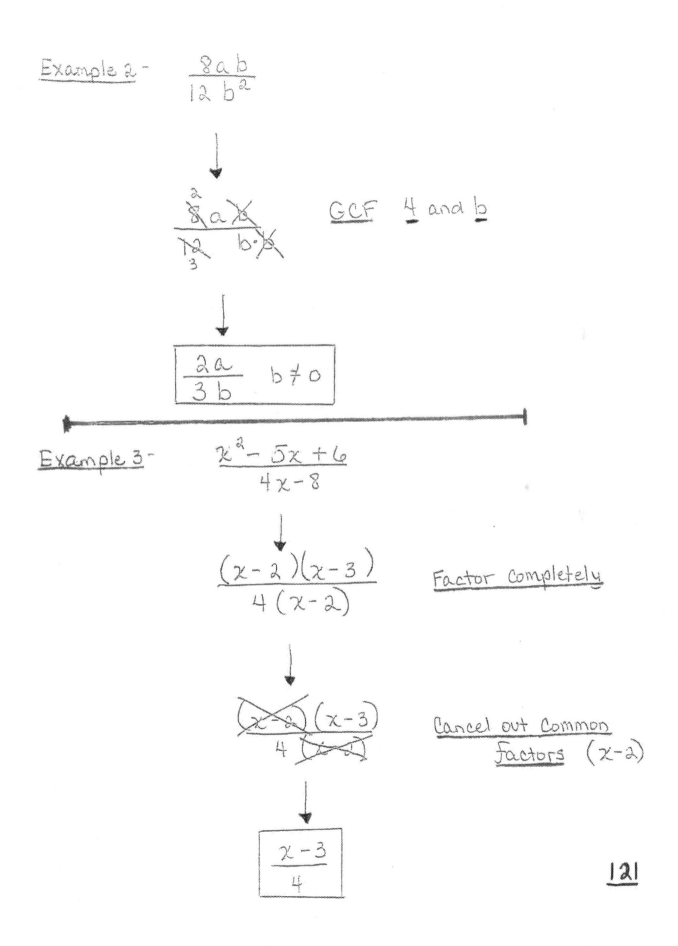

$\dfrac{\overset{2}{\cancel{8}}a\cancel{b}}{\underset{3}{\cancel{12}}\;\cancel{b \cdot b}}$

GCF $\underline{4}$ and $\underline{b}$

$$\boxed{\dfrac{2a}{3b} \quad b \neq 0}$$

Example 3 - $\dfrac{x^2 - 5x + 6}{4x - 8}$

$\dfrac{(x-2)(x-3)}{4(x-2)}$

Factor Completely

$\dfrac{\cancel{(x-2)}(x-3)}{4\,\cancel{(x-2)}}$

Cancel out Common Factors $(x-2)$

$$\boxed{\dfrac{x-3}{4}}$$

121

SIMPLIFYING A RATIONAL EXPRESSION

When simplifying a rational expression, simplify the numerator and denominator by factoring to simplest terms. Cancel out all common factors.

Example 1 -

$$\frac{x^2 - 5x + 4}{x - 1}$$ ← factor numerator

$$\downarrow$$

$$\frac{(x - 4)(x - 1)}{(x - 1)}$$

$$\downarrow$$

$$\frac{(x - 4)\cancel{(x-1)}}{\cancel{(x-1)}}$$ cancel out common factors

$$\downarrow$$

$$\frac{x - 4}{1} = \boxed{x - 4}$$

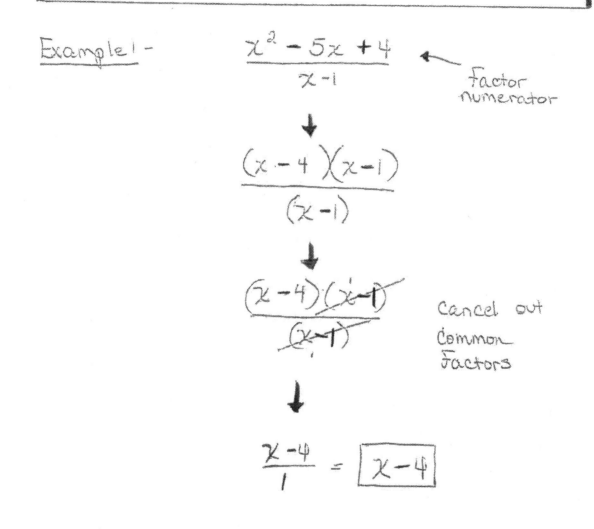

Example 2-

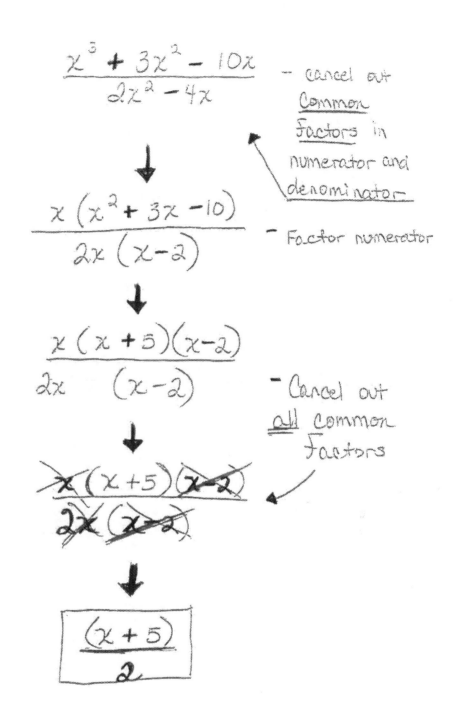

$$\frac{x^3 + 3x^2 - 10x}{2x^2 - 4x}$$

- cancel out common **factors** in numerator and denominator

↓

$$\frac{x(x^2 + 3x - 10)}{2x(x-2)}$$

- Factor numerator

↓

$$\frac{x(x+5)(x-2)}{2x \quad (x-2)}$$

- Cancel out **all** common factors

↓

$$\frac{x(x+5)(x-2)}{2x(x-2)}$$

↓

$$\boxed{\frac{(x+5)}{2}}$$

123

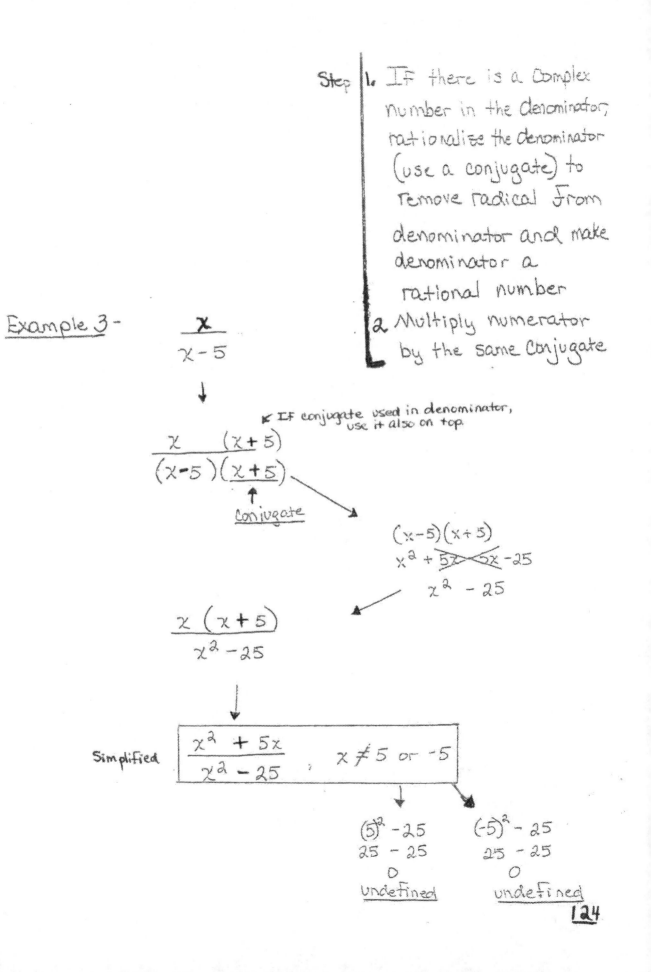

Step 1. IF there is a Complex number in the denominator, rationalize the denominator (use a conjugate) to remove radical from denominator and make denominator a rational number

2. Multiply numerator by the same Conjugate

Example 3 -

$$\frac{x}{x-5}$$

↓

* IF conjugate used in denominator, use it also on top

$$\frac{x}{(x-5)} \frac{(x+5)}{(x+5)}$$

Conjugate

$$(x-5)(x+5)$$
$$x^2 + 5x - 5x - 25$$
$$x^2 - 25$$

$$\frac{x(x+5)}{x^2-25}$$

↓

Simplified $\boxed{\dfrac{x^2 + 5x}{x^2 - 25} \quad , \quad x \neq 5 \text{ or } -5}$

$(5)^2 - 25$ $(-5)^2 - 25$
$25 - 25$ $25 - 25$
0 0
undefined   undefined

124

SIMPLIFYING RATIONAL EXPRESSIONS
WITH RADICALS

* Radicals may <u>not</u> be in the denominator of a rational expression

Example 1:

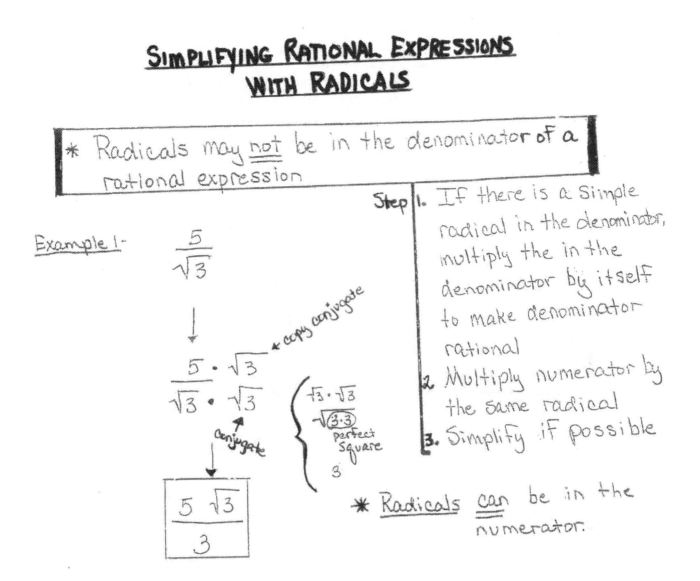

$$\frac{5}{\sqrt{3}}$$

$$\frac{5 \cdot \sqrt{3}}{\sqrt{3} \cdot \sqrt{3}}$$

← copy conjugate

conjugate

$$\begin{cases} \sqrt{3} \cdot \sqrt{3} \\ \sqrt{3 \cdot 3} \text{ perfect square} \\ 3 \end{cases}$$

$$\frac{5\sqrt{3}}{3}$$

Step 1. If there is a simple radical in the denominator, multiply the in the denominator by itself to make denominator rational
2. Multiply numerator by the same radical
3. Simplify if possible

* Radicals <u>can</u> be in the numerator.

SIMPLIFYING RATIONAL EXPRESSIONS WITH IMAGINARY NUMBERS

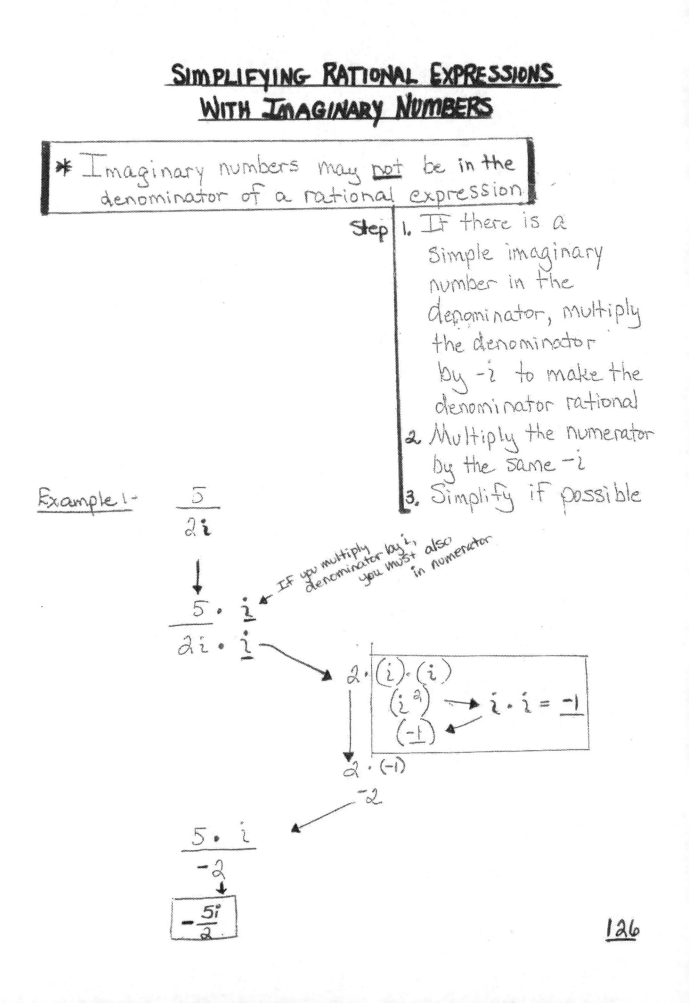

* Imaginary numbers may <u>not</u> be in the denominator of a rational expression.

Step 1. IF there is a simple imaginary number in the denominator, multiply the denominator by $-i$ to make the denominator rational

2. Multiply the numerator by the same $-i$

3. Simplify if possible

Example 1-

$$\frac{5}{2i}$$

$$\downarrow$$

$$\frac{5 \cdot i}{2i \cdot i}$$

If you multiply denominator by i, you must also in numerator

$$2 \cdot (i) \cdot (i)$$
$$(i^2)$$
$$(-1)$$
$$\longrightarrow i \cdot i = \underline{-1}$$

$$2 \cdot (-1)$$
$$\downarrow$$
$$-2$$

$$\frac{5 \cdot i}{-2}$$

$$\downarrow$$

$$\boxed{-\frac{5i}{2}}$$

126

1. IF there is a complex number in the denominator, rationalize the denominator (use a conjugate) to remove the imaginary number from the denominator and make the denominator a rational number

2. Multiply numerator by the same conjugate

Example 2-

$$\frac{4 + 2i}{4 - 2i}$$

$\downarrow$

$$\frac{(4 + 2i)}{(4 - 2i)} \overset{\text{Copied}}{\underset{\uparrow \text{ Conjugate}}{\frac{(4 + 2i)}{(4 + 2i)}}}$$

$\searrow$

Denominator

$$(4 - 2i)(4 + 2i)$$

$$\overset{F}{16} + \cancel{8i} - \cancel{8i} \overset{L}{- 4(i^2)}$$

$$\downarrow$$

$$16 - 4(-1)$$

$$\downarrow$$

$$16 + 4$$

$$\frac{(4 + 2i)(4 + 2i)}{\boxed{20}} \longleftarrow \underline{20}$$

Numerator

$$(4 + 2i)(4 + 2i)$$

$$\overset{F}{16} + \overset{O}{8i} + \overset{I}{8i} + \overset{L}{4i^2}$$

$$16 + 16i - 4(-1)$$

$$16 + 16i - 4$$

$\longrightarrow$ $12 + 16i$ $\longrightarrow$ $\frac{12 + 16i}{\boxed{20}}$ $\longrightarrow$ $\frac{\overset{3}{\cancel{12}} + \overset{4}{\cancel{16}}i}{\underset{5}{\cancel{20}}}$ $\longrightarrow$ $\boxed{\frac{3 + 4i}{5}}$

reduce by 4

121

COMPOSITE FUNCTIONS

Basically solving 2 equations - Solve innermost equation first, then use that solution to solve the outside equation (Solve backwards)

Example 1 - $f(x) = 3x - 2$ $g(x) = x^2 + 1$

Find $f(g(3))$

↓   Solve $g(3)$ First

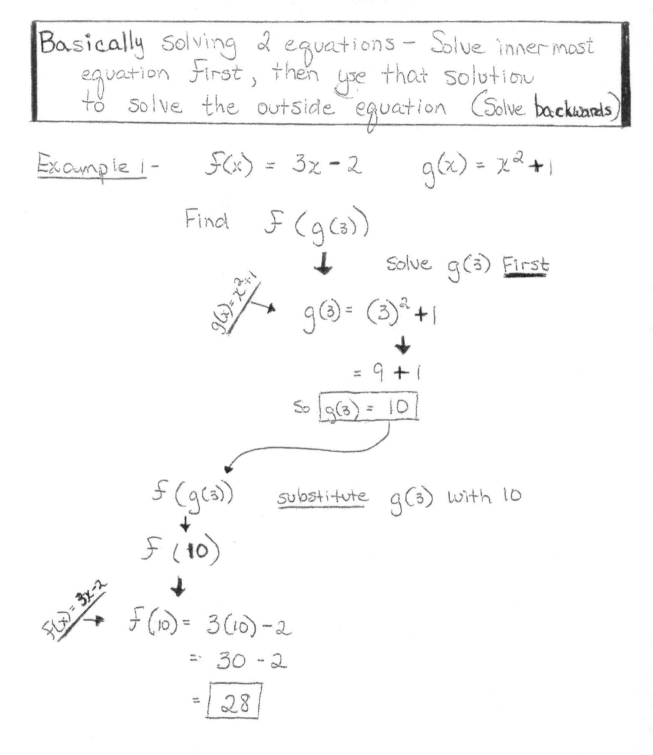

$g(3) = x^2 + 1$ → $g(3) = (3)^2 + 1$

↓

$= 9 + 1$

So $\boxed{g(3) = 10}$

$f(g(3))$ substitute $g(3)$ with 10

↓

$f(10)$

↓

$f(x) = 3x-2$ → $f(10) = 3(10) - 2$

$= 30 - 2$

$= \boxed{28}$

Example 2 - $f(x) = 3x-2$ $g(x) = x^2 + 1$

* This is the same
type problem as
Example 1 but
written in a
different form.

Find $(f \circ g)(-2)$

Solve backward
by substituting
(-2) in for x
in $g(x)$
equation

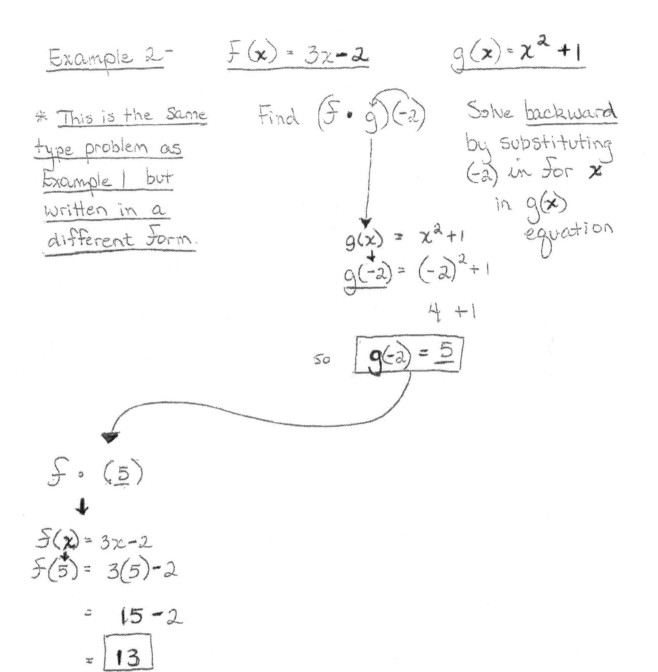

$g(x) = x^2 + 1$

$g(-2) = (-2)^2 + 1$

$4 + 1$

so $\boxed{g(-2) = 5}$

$f \circ (5)$

$f(x) = 3x-2$
$f(5) = 3(5)-2$

$= 15-2$

$= \boxed{13}$

INVERSE FUNCTIONS

Reverse the $\underline{x}$ and $\underline{y}$; then, solve for y.

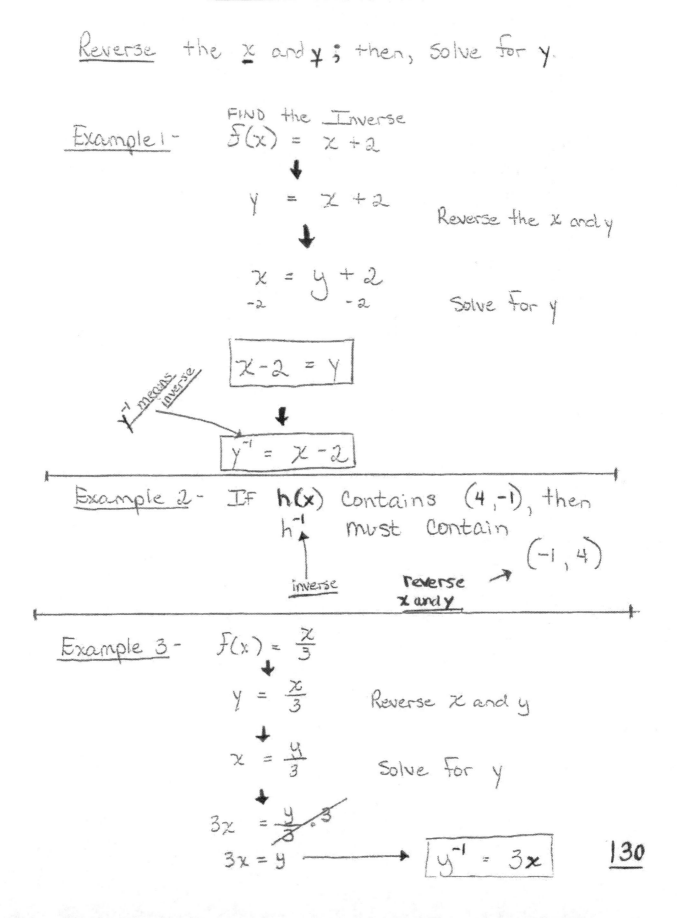

Example 1 -   FIND the Inverse
$$f(x) = x + 2$$

$$y = x + 2$$

Reverse the x and y

$$x = y + 2$$
$$-2 \qquad\qquad -2$$

Solve for y

$$\boxed{x - 2 = y}$$

y^{-1} means inverse

$$\boxed{y^{-1} = x - 2}$$

Example 2 -   IF $h(x)$ contains $(4, -1)$, then
h^{-1} must contain
$(-1, 4)$

inverse

reverse
x and y

Example 3 - $f(x) = \dfrac{x}{3}$

$$y = \dfrac{x}{3}$$

Reverse x and y

$$x = \dfrac{y}{3}$$

Solve for y

$$3x = \dfrac{y}{\cancel{3}} \cancel{\cdot 3}$$

$$3x = y \longrightarrow \boxed{y^{-1} = 3x}$$

130

DOMAIN AND RANGE OF A FUNCTION

Domain → x-values   Range → y-values

Example 1-   What is the domain?

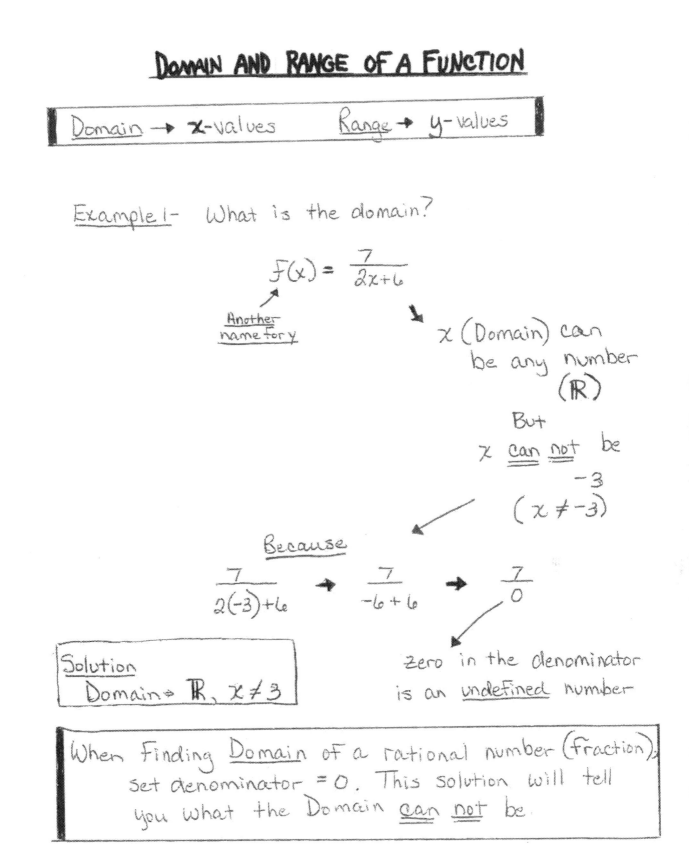

$$f(x) = \frac{7}{2x+6}$$

Another
name for y

x (Domain) can
be any number
$(\mathbb{R})$

But
x <u>can</u> <u>not</u> be
-3
$(x \neq -3)$

<u>Because</u>

$$\frac{7}{2(-3)+6} \rightarrow \frac{7}{-6+6} \rightarrow \frac{7}{0}$$

zero in the denominator
is an <u>undefined</u> number

Solution
Domain→ $\mathbb{R}$, $x \neq 3$

When finding <u>Domain</u> of a rational number (fraction),
set denominator $= 0$. This solution will tell
you what the Domain <u>can</u> <u>not</u> be.

Example 2 → What is the domain?

$$f(x) = \sqrt{x-7}$$

Another name for y

Radicals can be any number ≥ 0
(Radicals can not be negative)

$$\sqrt{x-7}$$

↓

$$x - 7 \geq 0$$

$$+7 \qquad +7$$

↓

All Real Numbers $x \geq 7$

Domain $\mathbb{R} \geq 7$

Example 3 - What is Range?

$$f(x) = \sqrt{x-7} + 2$$

Another name
for y

- If Domain ≥ 7, substitute
  smallest Domain (7)
  into equation and solve for y

$$y = \sqrt{(7)-7} + 2$$

$$\sqrt{0} + 2$$

$$0 + 2$$

$$\boxed{2}$$

So $f(x) \geq 2$
(Range) y $\geq$ 2

132

DIRECT VARIATION

$y = \underset{\uparrow}{\underline{k}}x$

Constant

- y varies directly as x
- y is directly proportional to x
- always linear (line)
- graph always runs through origin (0,0)

- As x changes, y changes at the same rate.

Example 1 -

$y = kx$
$y = \underline{2}x$
$\downarrow$

k = 2 (rate of change)

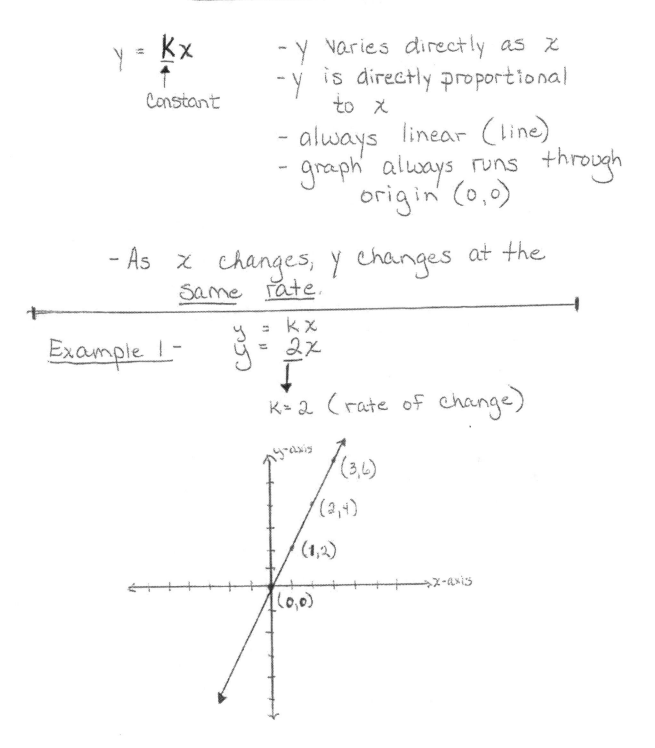

133

INVERSE VARIATION

$y = \dfrac{k}{x}$

- y varies inversely as x
- y is inversely proportional to x
- an increase in one variable results in a decrease in the other (x and y)
- K is some constant

Example 1 - $y = \dfrac{k}{x}$ if $k = 12$

x	y
-4	-3
-3	-4
-2	-6
-1	-12
0	X
1	12
2	6
3	4
4	3

← undefined $x \neq 0$

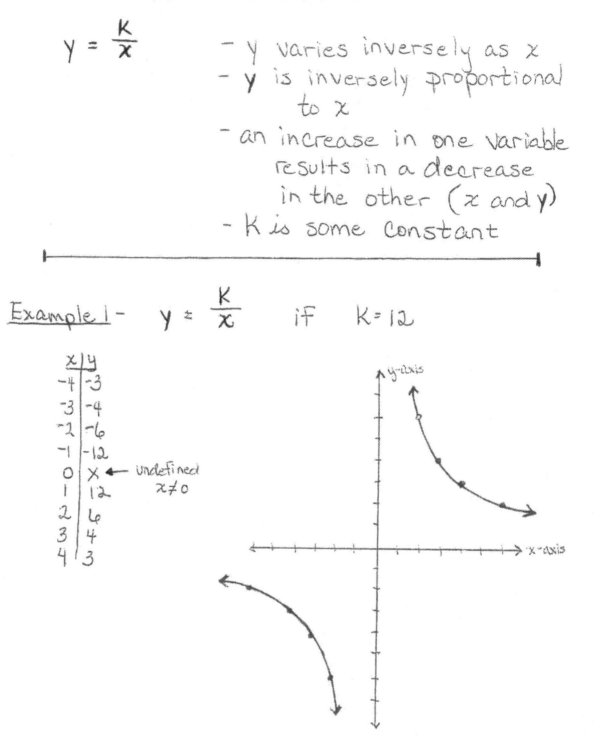

LOGARITHMS

Logarithms are exponent inverses

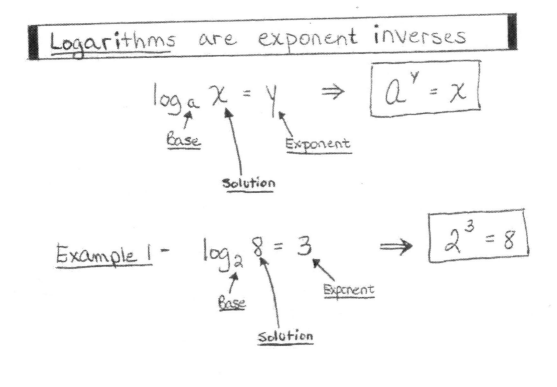

$$\log_a x = y \Rightarrow \boxed{a^y = x}$$

Base — Solution — Exponent

Example 1 - $\log_2 8 = 3 \Rightarrow \boxed{2^3 = 8}$

Base — Solution — Exponent

Example 2 - $\log_2 16 = x$

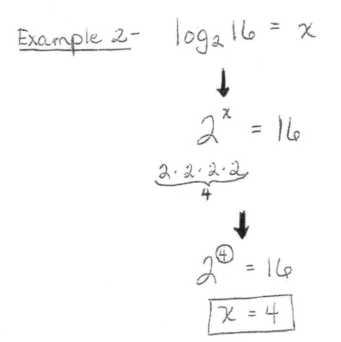

$$2^x = 16$$

$$\underbrace{2 \cdot 2 \cdot 2 \cdot 2}_{4}$$

$$2^{\textcircled{4}} = 16$$

$$\boxed{x = 4}$$

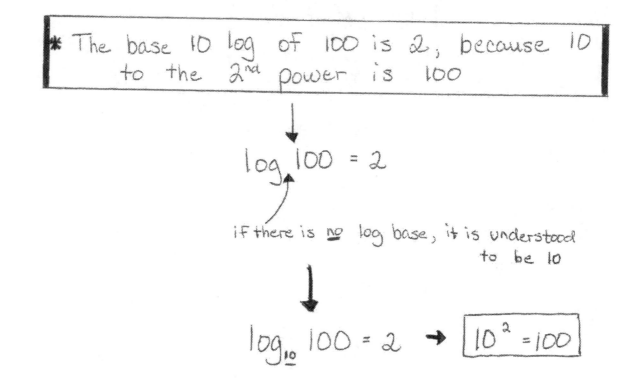

* The base 10 log of 100 is 2, because 10 to the 2nd power is 100

$$\log 100 = 2$$

if there is <u>no</u> log base, it is understood to be 10

$$\log_{10} 100 = 2 \quad \rightarrow \quad \boxed{10^2 = 100}$$

ADDING AND SUBTRACTING LOGARITHMS

When adding logs with <u>like</u> bases, <u>multiply</u> the terms

Rule → $\log_b x + \log_b y$

$$\log_b (x \cdot y)$$

Example 1 - $\log_2 4 + \log_2 8$

$$\log_2 (4 \cdot 8)$$

$$\boxed{\log_2 32}$$

When subtracting logs with <u>like</u> bases, <u>divide</u> the terms

Rule → $\log_b x - \log_b y$

$$\log_b \left(\frac{x}{y}\right)$$

Example 2 - $\log_2 8 - \log_2 4$

$$\log_2 \left(\frac{8}{4}\right)$$

$$\boxed{\log_2 (2)}$$

$$\boxed{\text{A } \underline{\text{coeffecient}} \text{ of a log is the } \underline{\text{exponent}} \text{ of the term}}$$

Rule → $2 \log_a x$

↓

$\log_a x^2$

Example 3 - $2 \log_a x - 3 \log_a y$

↓

$\log_a \left(\dfrac{x^2}{y^3} \right)$

RATE, RATIO AND PROPORTION

> Rate is a comparison of two quantities with different units

Example 1 — A car traveling 110 miles in 2 hours is moving at a rate of

$$\frac{110 \text{ miles}}{2 \text{ hours}} \quad \text{or} \quad 55 \text{ miles/hour}$$

> Ratios are comparing two unlike quantities — reduce if possible.

Example 2 — There are 5 girls and 8 boys in a math class. What is the ratio of girls to boys?

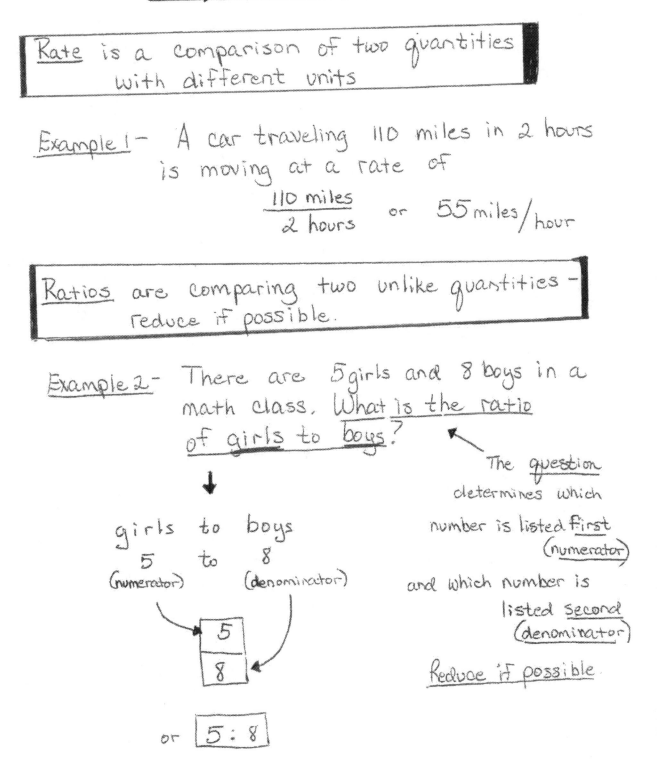

girls   to   boys
  5   to   8
(numerator)   (denominator)

$$\frac{5}{8}$$

The question determines which number is listed first (numerator) and which number is listed second (denominator)

Reduce if possible.

or 5 : 8

PYTHAGOREAN THEOREM

Pythagorean theorem is used to find the missing side of any right triangle if given 2 sides

Pythagorean Theorem → $a^2 + b^2 = c^2$

2 sides (Legs)

Hypotenuse (longest side)

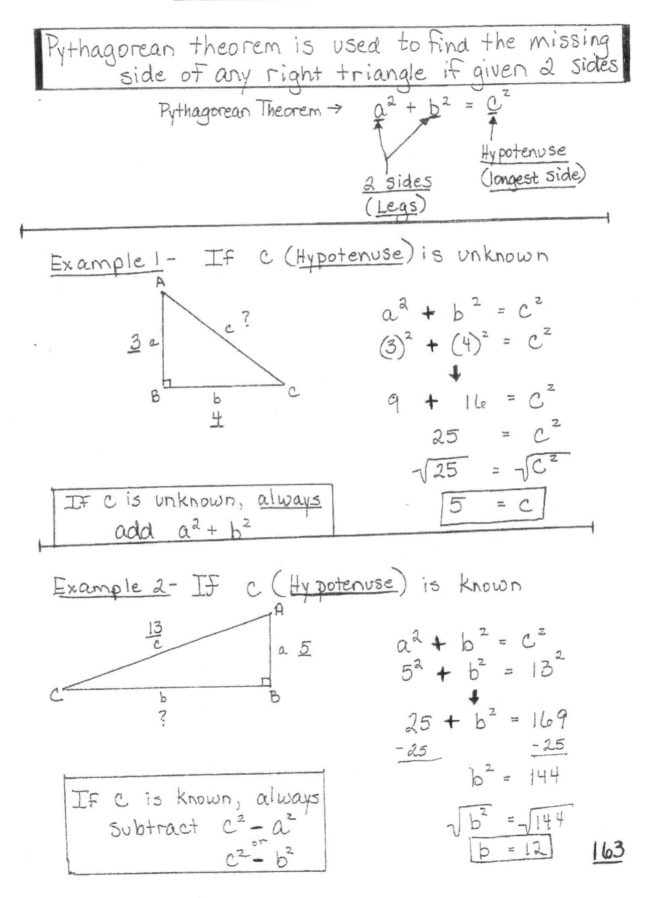

Example 1- If c (Hypotenuse) is unknown

$$a^2 + b^2 = c^2$$
$$(3)^2 + (4)^2 = c^2$$
$$9 + 16 = c^2$$
$$25 = c^2$$
$$\sqrt{25} = \sqrt{c^2}$$

If c is unknown, always add $a^2 + b^2$

$$5 = c$$

Example 2- If c (Hypotenuse) is known

$$a^2 + b^2 = c^2$$
$$5^2 + b^2 = 13^2$$
$$25 + b^2 = 169$$
$$-25 \qquad -25$$
$$b^2 = 144$$
$$\sqrt{b^2} = \sqrt{144}$$
$$b = 12$$

If c is known, always subtract $c^2 - a^2$ or $c^2 - b^2$

163

SPECIAL RIGHT TRIANGLES

45-45-90 (Isosceles Right Triangle or Diagonal of a Square)

Example 1 -

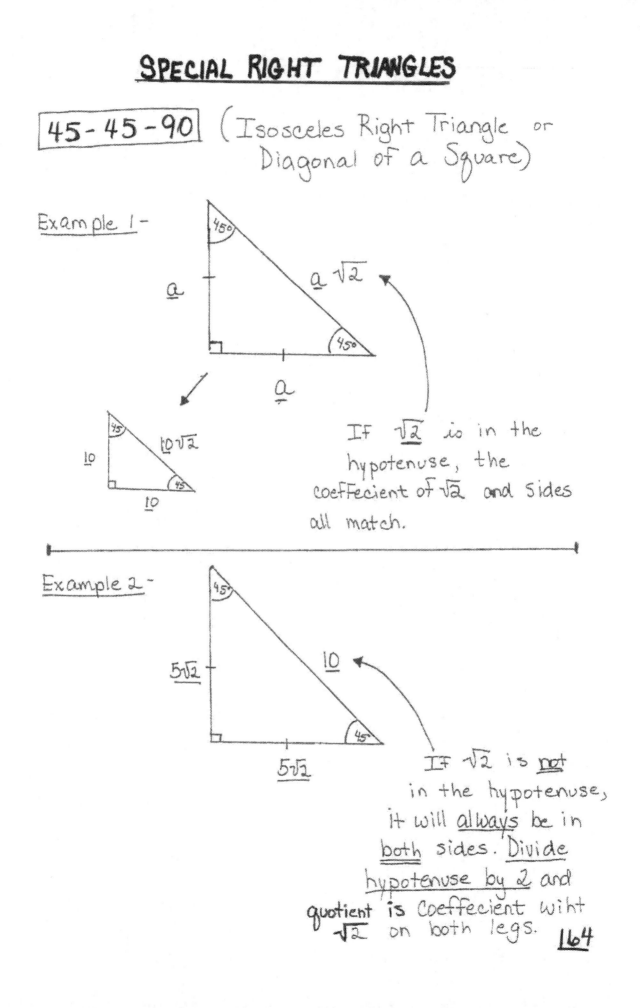

If $\sqrt{2}$ is in the hypotenuse, the coeffecient of $\sqrt{2}$ and sides all match.

Example 2 -

If $\sqrt{2}$ is not in the hypotenuse, it will always be in both sides. Divide hypotenuse by 2 and quotient is coeffecient wiht $\sqrt{2}$ on both legs.

164

30-60-90 (Half an Equalateral Triangle)

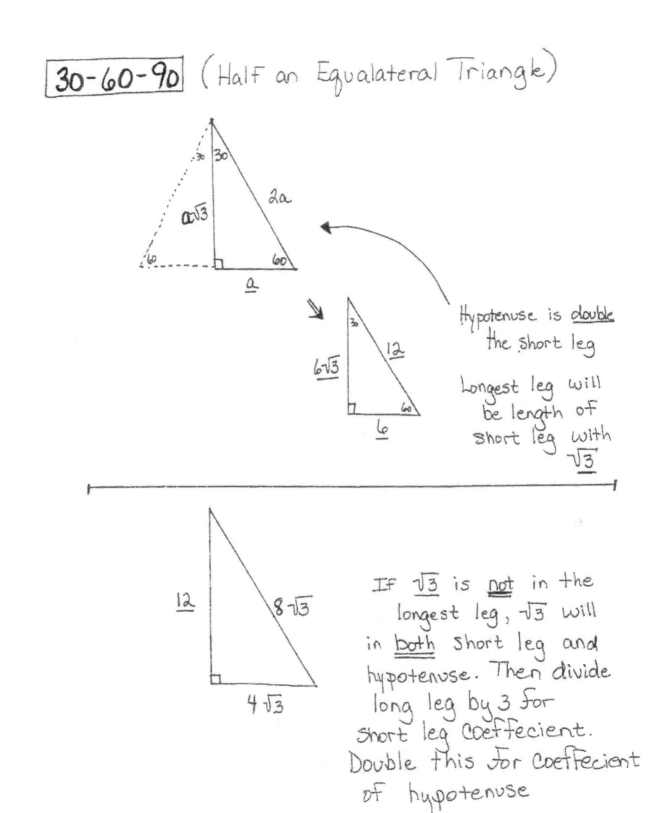

Hypotenuse is <u>double</u> the short leg

Longest leg will be length of short leg with $\sqrt{3}$

If $\sqrt{3}$ is <u>not</u> in the longest leg, $\sqrt{3}$ will in <u>both</u> short leg and hypotenuse. Then divide long leg by 3 for short leg coeffecient. Double this for coeffecient of hypotenuse

SIMILAR FIGURES

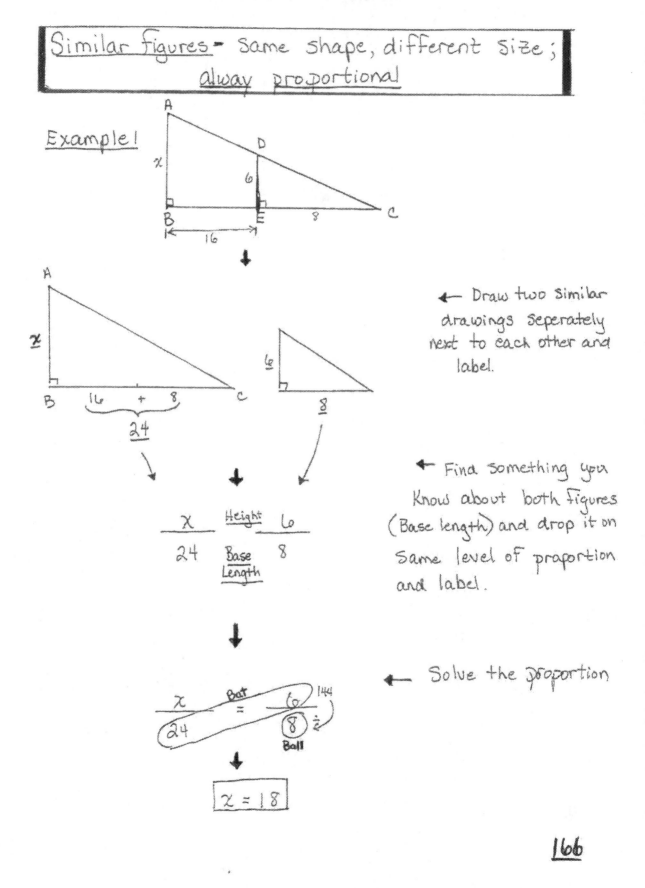

Similar figures - Same shape, different size; alway proportional

Example1

← Draw two similar drawings seperately next to each other and label.

← Find something you know about both figures (Base length) and drop it on same level of proportion and label.

$$\frac{x}{24} \quad \frac{\text{Height}}{\text{Base Length}} \quad \frac{6}{8}$$

← Solve the proportion

$$\frac{x}{24} \overset{\text{Bat}}{=} \frac{6}{8} \quad \frac{144}{\div}$$

$$x = 18$$

Lengths in similar figures are proportional, but **all** Corresponding angles are <u>Congruent</u>

Example 2 -

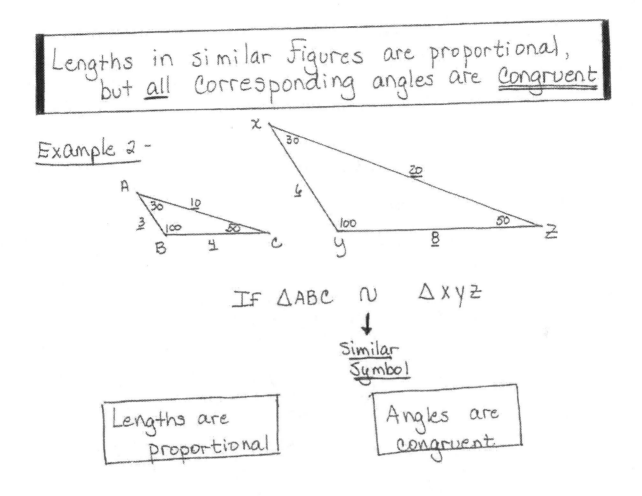

IF △ABC ∿ △XYZ

↓

<u>Similar</u>
<u>Symbol</u>

Lengths are proportional

Angles are congruent

DEGREES IN A POLYGON

Polygon- 3 or more sides, straight line segments that are closed.

Polygons are classified according to the number of line segments used as sides

number of sides	Name	Number of total degrees
3	Triangle	180°
4	Quadrilateral	360°
5	Pentagon	540°
6	Hexagon	720°
7	Heptagon	900°
8	Octagon	1080°
9	Nonagon	1260°
10	Decagon	1440°

Formula to find degrees in any polygon.

$$\text{Degrees} = 180 (n - 2)$$

$\downarrow$

number of sides

Example 1- How many degrees in a 12- sided polygon?

$$D = 180 (12 - 2)$$
$$180 (10)$$
$$\boxed{D = 1800}$$

168

PERIMETER

Perimeter - distance around the outside of any polygon

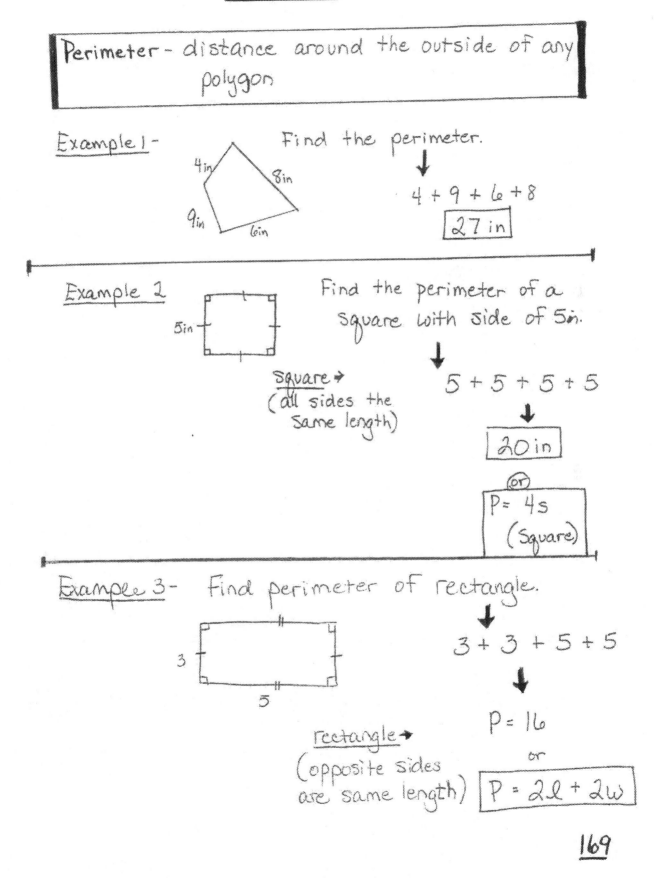

Example 1 -   Find the perimeter.

4in   8in   9in   6in

4 + 9 + 6 + 8

$\boxed{27 \text{ in}}$

Example 2   Find the perimeter of a square with side of 5in.

5in

square →
(all sides the
same length)

5 + 5 + 5 + 5

$\boxed{20 \text{ in}}$

or

P = 4s
(square)

Example 3 -   Find perimeter of rectangle.

3   5

3 + 3 + 5 + 5

P = 16

rectangle →
(opposite sides
are same length)

or

$\boxed{P = 2l + 2w}$

169

AREA

Area - always expressed in square units

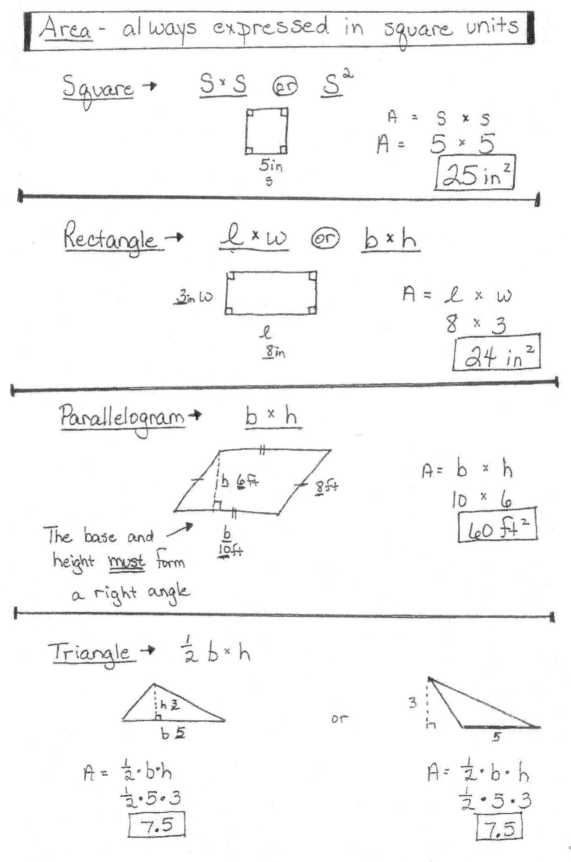

Square → $S \times S$ or S^2

$A = S \times S$
$A = 5 \times 5$
$\boxed{25 \, in^2}$

5in
s

Rectangle → $\ell \times w$ or $b \times h$

3in W

ℓ
8in

$A = \ell \times w$
8×3
$\boxed{24 \, in^2}$

Parallelogram → $b \times h$

h 6ft   8ft

b
10ft

The base and height __must__ form a right angle

$A = b \times h$
10×6
$\boxed{60 \, ft^2}$

Triangle → $\frac{1}{2} b \times h$

h 3
b 5

or

3
h
5

$A = \frac{1}{2} \cdot b \cdot h$
$\frac{1}{2} \cdot 5 \cdot 3$
$\boxed{7.5}$

$A = \frac{1}{2} \cdot b \cdot h$
$\frac{1}{2} \cdot 5 \cdot 3$
$\boxed{7.5}$

170

Trapezoid → $\frac{1}{2}(b_1 + b_2)h$

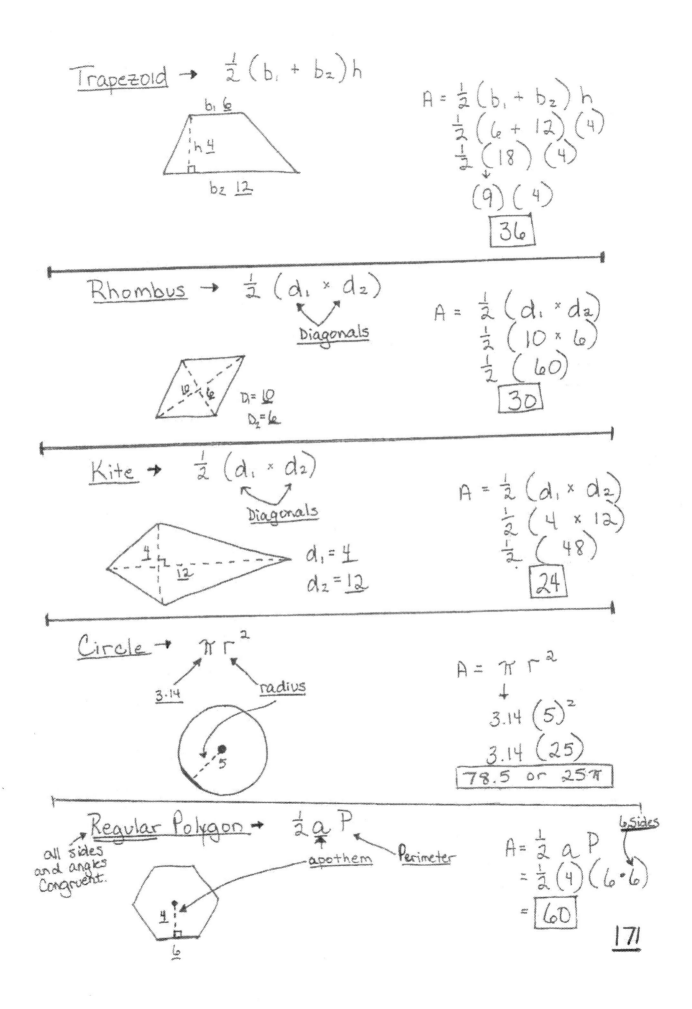

b_1 6
h 4
b_2 12

$A = \frac{1}{2}(b_1 + b_2)h$

$\frac{1}{2}(6 + 12)(4)$

$(18)(4)$

$(9)(4)$

$\boxed{36}$

Rhombus → $\frac{1}{2}(d_1 \times d_2)$

Diagonals

$D_1 = 10$
$D_2 = 6$

$A = \frac{1}{2}(d_1 \times d_2)$

$\frac{1}{2}(10 \times 6)$

$\frac{1}{2}(60)$

$\boxed{30}$

Kite → $\frac{1}{2}(d_1 \times d_2)$

Diagonals

$d_1 = 4$
$d_2 = 12$

$A = \frac{1}{2}(d_1 \times d_2)$

$\frac{1}{2}(4 \times 12)$

$\frac{1}{2}(48)$

$\boxed{24}$

Circle → πr^2

3.14   radius

5

$A = \pi r^2$

$3.14(5)^2$

$3.14(25)$

$\boxed{78.5 \text{ or } 25\pi}$

Regular Polygon → $\frac{1}{2}aP$

all sides and angles congruent.

apothem   Perimeter

6 sides

4

6

$A = \frac{1}{2}aP$

$= \frac{1}{2}(4)(6 \cdot 6)$

$= \boxed{60}$

171

VOLUME

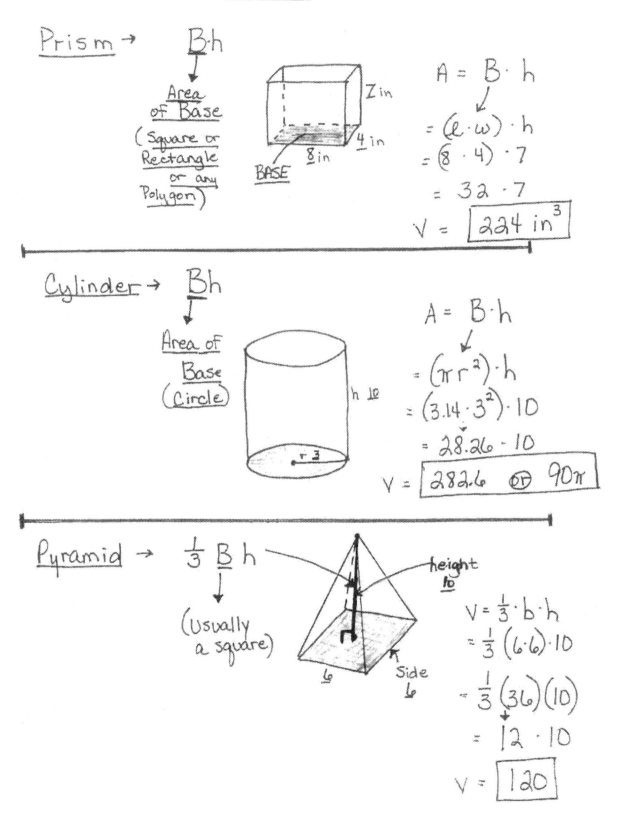

Prism → $\underline{B \cdot h}$

↓

$\underline{\text{Area}}$
of $\underline{\text{Base}}$

(Square or
Rectangle
or any
Polygon)

BASE

2 in

4 in

8 in

$A = B \cdot h$

$= (\ell \cdot w) \cdot h$

$= (8 \cdot 4) \cdot 7$

$= 32 \cdot 7$

$V = \boxed{224 \text{ in}^3}$

Cylinder → $\underline{Bh}$

↓

$\underline{\text{Area of}}$
$\underline{\text{Base}}$
(Circle)

h 10

r 3

$A = B \cdot h$

$= (\pi r^2) \cdot h$

$= (3.14 \cdot 3^2) \cdot 10$

$= 28.26 \cdot 10$

$V = \boxed{282.6 \quad \text{or} \quad 90\pi}$

Pyramid → $\frac{1}{3} \underline{B} h$

↓

(usually
a square)

height
10

Side
6

6

$V = \frac{1}{3} \cdot b \cdot h$

$= \frac{1}{3} (6 \cdot 6) \cdot 10$

$= \frac{1}{3} (36)(10)$

$= 12 \cdot 10$

$V = \boxed{120}$

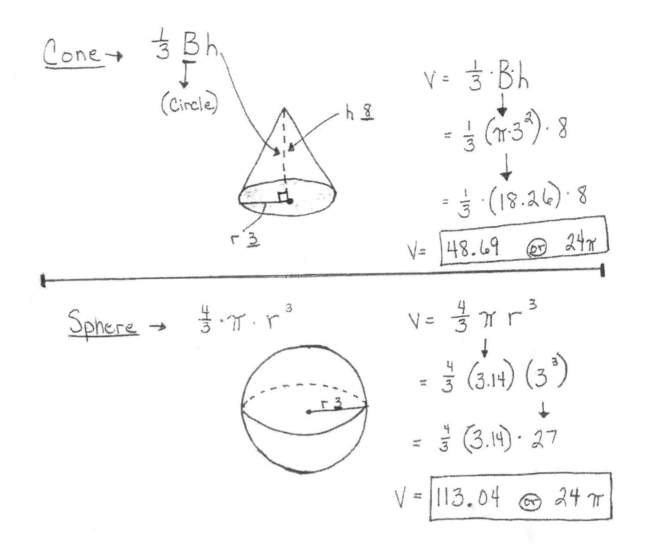

Cone → $\frac{1}{3} B h$

↓

(Circle)

h 8

r 3

$V = \frac{1}{3} \cdot B \cdot h$

$= \frac{1}{3} (\pi \cdot 3^2) \cdot 8$

$= \frac{1}{3} \cdot (18.26) \cdot 8$

$V =$ 48.69 or 24π

Sphere → $\frac{4}{3} \cdot \pi \cdot r^3$

r 3

$V = \frac{4}{3} \pi r^3$

$= \frac{4}{3} (3.14) (3^3)$

$= \frac{4}{3} (3.14) \cdot 27$

$V =$ 113.04 or 24π

173

CIRCLE BASICS

Chord - Line segment that runs from one edge point to another edge point on a circle. A chord <u>does</u> <u>not</u> have to touch the center of the circle (<u>Diameter is the longest chord</u>)

Diameter - Line segment that runs from one edge point to another edge point of a circle, but <u>Must</u> run through the Center.

Radius - Line segment that runs from center of circle to a point on the outside edge. (<u>Radius is ½ the diameter</u>)

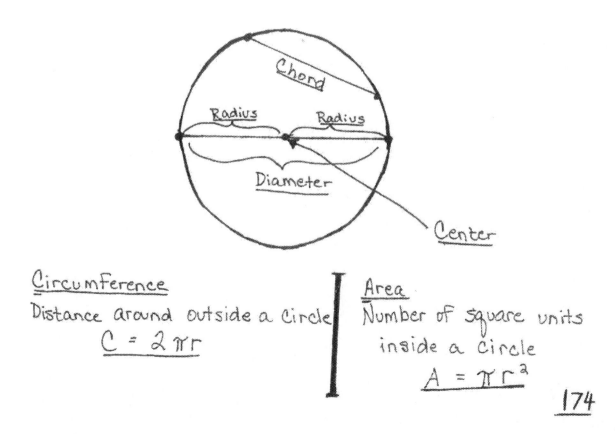

<u>Circumference</u>
Distance around outside a circle
$$C = 2\pi r$$

<u>Area</u>
Number of square units inside a circle
$$A = \pi r^2$$

174

FORMULA OF A CIRCLE

When center is $(0,0)$ → $x^2 + y^2 = r^2$
on a graph

Example 1 - $\quad x^2 + y^2 = 4^2$

center → $(0,0)$ Radius → 4

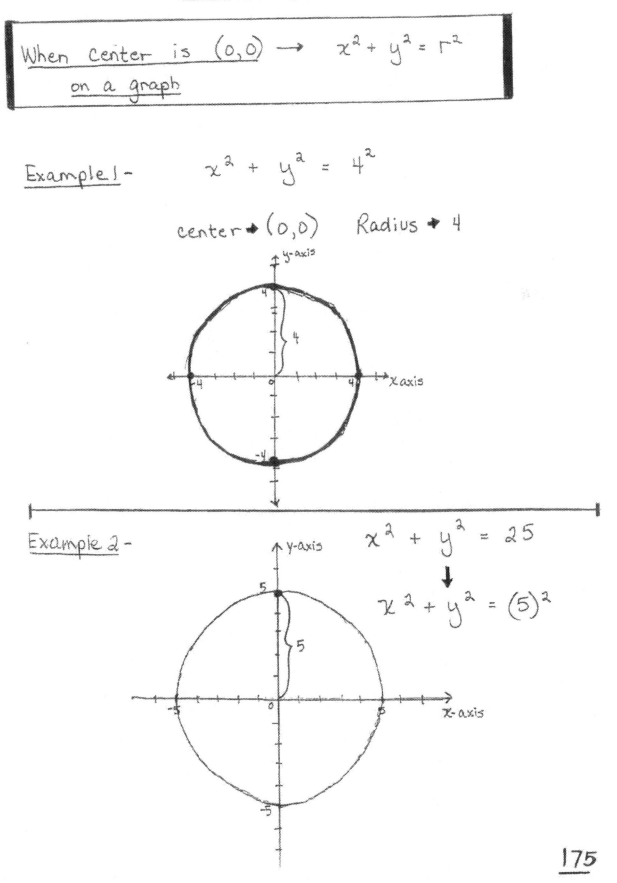

Example 2 - $\quad x^2 + y^2 = 25$

$$\downarrow$$

$$x^2 + y^2 = (5)^2$$

$$\boxed{\text{When } \underline{\text{center}} \text{ is } \underline{(h,K)} \rightarrow (x-h)^2 + (y-k)^2 = r^2}$$
$$\text{on a } \underline{\text{graph}}$$

$\underline{\text{Example 3}}$ — $\qquad (x+2)^2 + (y-3)^2 = 1$

$\downarrow$

$$\left(x-(-2)\right)^2 + \left(y-(3)\right)^2 = (1)^2$$

$\qquad\qquad\qquad \downarrow \qquad\qquad\qquad \downarrow \qquad\qquad \downarrow$

$\qquad\qquad\quad \underline{h=-2} \qquad\qquad \underline{K=3} \qquad\quad r=1$

$\qquad\qquad\qquad\qquad\qquad (-2,3) \qquad\qquad \underline{\text{Radius}} = 1$

$\qquad\qquad\qquad\qquad\qquad \underline{\text{Center}}$

radius = 1

Center (-2,3)

g-axis

x-axis